MASTERING ALGEBRA

An Axiomatic Approach

Second Edition

by

Roger W. Oster

Oster

Cover Background: Courtesy NASA-JPL (Hubble: Mystic Mountain)

Preface to the Second Edition

This new edition includes many more practice problems, but only enough to better help you tackle and absorb the algebraic topic at hand. Some are easy, some are intermediate, and some will require your full, focused attention. I have also made some minor corrections to the original text to accommodate the practice problems, and I have expanded the book to include some important, more advanced algebraic topics. Logarithms, factoring, solving equalities and inequalities, functions, graphing, and complex numbers have all been included in this new edition. I have also included a complete answers-to-problems section at the end of the book, and the cover design has been changed as well (which has more to it than meets the eye).

Becoming a master of any subject, whether it be watercolors, poetry, musical composition or mathematics, requires work, but the rewards immensely outweigh the effort. However, that doesn't mean you have to take the hard and zigzaggy road to success; there *are* straight roads that open up to vast illuminated hillsides which allow you to see the entire journey and destination in one revelatory flash; this book is one of those roads. It was designed for you to take advantage of your innate creative and logical abilities, and for you to effectively utilize the repetitive characteristics—the patterns, if you will—of algebra to not only master algebra, but to understand the workings behind much of mathematics in general. It was designed in many ways to let algebra, itself, do much of the hard work for you. I fully trust that you will find this book an enjoyable and enlightening read.

This fine book has been four years in the making and has been extensively edited for format, ease of reading, spelling, grammar and mathematical correctness, but it is a math book, and all math books are prone to mistakes—it is the nature of subject. With that said, if you would like to be aware of any updates or addendums for this book or if you would just like to drop a line or two, then please visit my website at RogerOster.com or email me at roger@rogeroster.com. I welcome and look forward to all your thoughts, questions and comments and promise to reply as soon as possible.

Roger W. Oster
August 2018

Contents

Introduction

Scratches on bone, such is the inglorious dawn of mathematics. It is a crude awakening by numbers' standards because these digital processors, by dint of their celestial existence and curious structure, have been ruling the movements of the stars, the planets and the galaxies since the birth of the Universe. For how else are we to explain a universe so keenly bound to and characterized by the axiomatic structures discovered by Euclid, Galileo, Newton, and Einstein? Surely, all the mysteries of the Universe would be revealed if we could only fathom the deep, "atomic" structure of numbers.

As relative newcomers to the study of this structure, however, we will neither go so far back as counting scratches on bone, nor so far forward as performing relativistic calculus. We will be examining the structure of numbers via the logical consequences of the axioms, theorems and definitions of modern algebra—i.e., by way of the *axiomatic method*—and per chance, by and by, gain a deeper appreciation for the methods and practice of all mathematics.

However, the axiomatic method is a radical departure from the traditional methods by which algebra is taught. Traditionally, algebra is taught through example and practice with strong reliance on the students' intuition about the properties of numbers. Audit just about any high school algebra course* and you will find that for each algebraic rule introduced—all too often via the tedium of lecture—the teacher will completely work two or three archetypical problems on the board, while addressing any student-questions with analogous arithmetical examples. For example, if the rule that students are having trouble with is the associative rule of addition, that is $a+(b+c)=(a+b)+c$, then the teacher may go on to explain the rule (and its reasonableness) with, say, $1+(2+3)=(1+2)+3$ and rely on the students' intuition to understand how the analogy between the two equations allows them to apply this arithmetical property of numbers to equations in algebra. Finally, at the end of each lesson the teacher will usually assign ten to twenty practice problems for homework.

This traditional teaching method, however, fails to connect with many students because it does not explain the purpose of the rules or their origins. To many students, the rules make no logical, cohesive sense—why was *this* or *that* rule chosen in the first place? Furthermore, if a student asks about the *why* of a rule, the teacher may, out of habit, time constraints or lack of confidence, "miss" the point of the question and proceed to explain the *how* of the rule instead—i.e., how the rule functions or is used as opposed to its usefulness or why the rule exists.

---

* The most popular textbook for which seems to be by Brown, Dolciani, Sorgenfrey and Cole (1997); *Algebra: Structure and Method (Books 1 & 2)*; Boston; McDougal Littel; (all editions)

On the other hand, if the teacher is thoroughly conversant with the axioms and does understand the point of the question but has never discussed the axioms of algebra with the class, then the methods available to the teacher to answer the question in any reasonably concise manner that would diminish or remove the student's confusion will be limited. It is only through a rather in-depth (but not necessarily time intensive) discussion about the axioms and their logical consequences that the *why* of an algebraic rule can be appreciably answered.

It is equally important, however, to learn the axiomatic method, as such, for the sheer sake of it because it is essential for critical thinking: Euclid (axioms of geometry), Descartes ("I think, therefore I am"), Thomas Jefferson ("... all men are created equal"), Newton (axioms of force) and Einstein (axioms of relativity) all relied on only a few axioms to deduce a picture of the world that, for better or worse, has changed the course of human history—a picture that can be denied only by the facts of nature, the human condition or the death of reason.

Remarkably however, the axiomatic method has made only dawdling advances beyond the very familiar but many-decades-old course of Euclidean geometry. One can postulate that the reason for this slow and procrastinated acceptance into the curriculums is due, at least in part, to the two historically conflicting methods mathematicians have used, and believe should be used, in the pursuit of truth in mathematics: intuition and logic, which in modern pedagogical terminology have been dubbed the different *styles* of learning or understanding. (We may also consider these styles as right-brain-intuition and left-brain-logic.)

The great French mathematician Henri Poincaré (1854 - 1912) first pointed out these two "styles" of learning/understanding as the psychological attributes belonging to the two distinct types of mathematicians*: the intuitionalists and the (mathematical) logicians, neither of whom, because it is intrinsic to their very nature, can understand (or certainly not easily understand) mathematics from the perspective of the other, and neither of whom are *readily* willing to accept the other's critique of what establishes legitimate proof.

This propensity of each person to lean more heavily towards one or the other perspective is also a primary reason why many students, depending on their perspective, have difficulty learning algebra (and much of mathematics in general), which, as we shall see, is taught overwhelmingly from an intuitive perspective. However, much of the incertitude and trepidation that students have with algebra can be tempered with a better understanding of what each of these perspectives are. Thus, a brief discussion about each is in order.

---

* Henri Poincaré published *Intuition and Logic in Mathematics* as part of *La Valeur de la Science* in 1905. (See Bib. for a more detailed reference)

The intuitionalists, the geometers as they are often called, understand mathematics through meaning and intuition—an *internal* sense about mathematics. In their minds, mathematics must be about something either conceptualizable or constructible, and they have an innate ability to use their "gut feelings" based on the images they can mentally construct from mathematical concepts. *Proof* to an intuitionalist is about obviousness (but that is not at all to say that logic isn't used by intuitionalists).

Contrastingly, the logicians, a.k.a. the algebraists, understand mathematics through logic—an *external* gathering of information acted upon by reason. In their minds, mathematics is about the logical connections between abstract concepts and words, and they have an innate ability to make those connections. *Proof* to a logician is about the final statement in a sequence of logically derived statements, the initial motivation of which are the "first" statements or axioms.

The distinction between the two types can be better explained through a comparison of the modern teaching methods used in geometry with those used in algebra (never mind the fact that the methods used in each are, and have been for some time, somewhat inverted given the nature of these two subjects). Geometry, the more intuitive subject because it deals with images, is taught with a greater reliance on words, logic and the consequences of its axioms than it is on intuition, whereas algebra, which is considered the more logical subject, is taught with a greater reliance on intuition.

For example, the idea of congruency in geometry is easily understood, *intuitively*, through the concept of transformations, where two geometrical figures are considered congruent if one can rotate and/or reflect the image of (i.e., transform in one's mind) one of the figures in such a way that it can be shown to be identical, both in size and shape, to the other.*

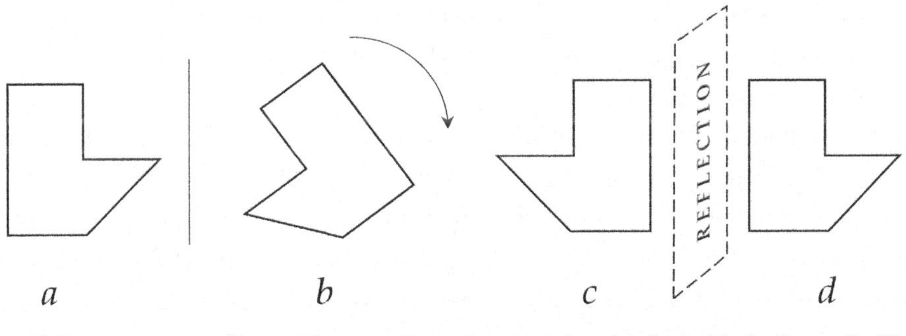

Figure b is congruent to figure a because it can be rotated and reflected to look exactly like a.

In contrast however, students of modern geometry are required to prove that two figures are congruent by showing that all of the angles and line segments (each at a time) in one figure are exactly like—congruent to—those of

---

*Many popular high school geometry texts do not introduce the spatial aspects of transformations until the last chapter, which means most of these concepts are usually omitted. (e.g., Jurgensen, R. C., Brown, R. G., Jurgensen, J. W.; (2000) *Geometry*. McDougal Littell.)

the other figure, and that they are in a *corresponding* order, which means that the corresponding sides of each figure fall between their corresponding angles, and that the corresponding angles of each figure fall between their corresponding sides. This challenging definition of the concept of congruent figures, which is part of the axiomatic, logical, presentation of geometry, can be confusing without including the more intuitive concept of transformations.

Algebra, on the other hand, is taught overwhelmingly from the perspective of intuition: students are expected to know the general properties of numbers and arithmetic and how they relate to algebra, and how addition, multiplication, equality and inequality "function", and how numbers are organized; for example, students are expected to know and accept, *without proof*, that $-1<1$.

Furthermore, most students' intuition about algebra (and mathematics in general) has developed via their elementary education; thus, the reliability of their mathematical intuition is dependent upon the quality of that education. Logic and reason on the other hand, being more related to cause and effect and our everyday life experiences is mostly independent of one's mathematics education, although in this respect, mathematics does serve to strengthen one's reasoning skills.

For example, prior to Galileo most people, including many scientists, philosophers and intellectuals, believed that if you dropped a marble and a cannon ball from the same height, the cannon ball would hit the ground first. This was not a failure of logic and experience, but rather a failure of intuition—faulty beliefs connected to the weights of objects had influenced their fundamental notions.

Regardless of their differences in perspective, or in their concept of the nature of proof, both mathematicians would agree with the need for proof and logic in the pursuit of mathematical knowledge, and to my readers who are educators, we are obliged to accommodate both types of student-mathematicians—both styles of learning.

When teaching or learning mathematics, therefore, it is incumbent upon the teacher *and* student to strike a balance between both perspectives if the path to knowledge is to provide the greatest opportunities for learning. With that said it is interesting to note that whenever algebra is explained from an axiomatic, logical perspective, it answers all of the questions that students often wonder about—questions that cannot be satisfactorily answered based purely on intuition alone. For example, why is the product of two negative numbers a positive number?* Why are negative numbers less than positive numbers? Why does multiplication take precedence over addition? Why is zero times any

---

* By analogy, students are often told that if you do not *not-do* something, then you in fact *do* that thing; this, however, although an *intuitive* explanation, and only a tolerable one at best, for the *product* of two negatives, fails to explain how a double negative in English relates to *multiplication*.

number zero?* Why does the product of the square roots of two numbers equal the square root of the product of the two numbers?

Curiously, the step-by-step, mechanical solution of an algebra problem (barring any superfluous rules that teachers often needlessly introduce like the butterfly rule, cross multiplication, invert-and-multiply, Please-Excuse-My-Dear-Aunt-Sally, etc.) is a rendering of the logical application of the axioms and theorems anyway, but tragically students are not, for the most part, made aware of that and are therefore blind to the collection of skillfully selected axioms† that when loosed upon an inquisitive and logical mind gives rise to the incredibly complex structure known as algebra and elementary arithmetic. I suppose the obvious question to ask is "why are there no formal proofs in Algebra 2, and in a more explorative, 'hands-on' way, in Algebra 1?"

Certainly, a contributing factor to this "oversight" is that while many teachers are quite comfortable with demonstrating the mechanics of algebra on the blackboard and naming the algebraic rule they are applying and how it is used, many also lack the theoretical knowledge or training to explain the logical purpose of the axioms, their interconnectedness and the theorems derivable from them.‡

Additionally, the mathematical community has long-established algebra as an applied mathematics course (i.e., a means to an end) as opposed to a theoretical one (i.e., an end in itself); thus, mechanical proficiency and real-world problem solving have been the primary focus of algebra, while proof and (theoretical) understanding have been abandoned to the vagaries of each student's intuition.

Unfortunately, it is unlikely that things in the classroom will change anytime soon. Institutions are notoriously rigid, and the stubborn persistence of "hand" calculating numerical problems (which are more efficiently solved with calculators and computers), found throughout much of the mathematics curriculums around the world, could be drastically reduced (i.e. it would allow for the greater use of calculators and computers to solve these problems) once students thoroughly understand the deeper concepts of algebra.

Few books on the market, however, do anything more than just repeat the problematic format of high school and college textbooks; most, although they are replete with catchy fonts and flowery pages and may even profess to instill

---

* What does it mean to "multiply" a number zero times? Does it mean to multiply using nothing as the multiplier? Does it mean "don't multiply"? How can nothing be a multiplier? What does multiplication mean in this context? Students are faced with many, if not all, of the same philosophical dilemmas that have plagued mathematicians since the dawn of numbers: intuition is historical and generational, and it has its limits.

† Arguably the most important of which—the closure axioms—are, for whatever unfathomable reason, expelled to the footnotes in high school algebra if they are even mentioned at all.

‡ See Bib.: Diana White (2015).

enlightenment upon the "dummies" of the world, are mere facades designed around the belief that understanding comes from entertainment, regardless of how much entertainment may, in fact, play a role in education. Realistically, however, what is needed is a book that not only affords deep insights into algebra, but one that, in the paraphrased words of Albert Einstein, explains it in such a way that your grandmother could understand it.

This book was designed to cut through the fog of unanswered questions that have needlessly confused students of algebra for far too long. It was also designed as a down-to-earth, easy to read supplement to be considered alongside standard textbooks of high school algebra, and as a stand-alone, short, but powerful introduction to the methods and practice of algebra.

Those who may benefit from this book are (1) all those who have resigned themselves to believing that they will never be able to understand algebra, but would, nevertheless, like to, and to do so without performing reams of algebraic problems; (2) high school students who are looking to gain a better understanding and grade in algebra or calculus (*However, this book is not recommended if you have not, at least, satisfactorily completed Algebra I and a course of geometry.*); (3) college students who are preparing to take a course in Abstract or Linear Algebra; and (4) educators, whether they are gainfully employed school teachers, tutors, or parents who homeschool, because this book can be used as a guide to developing a deeper, student understanding of Algebra I, Algebra II, and the axiomatic method; it also hints at some creative activities to communicate the power and extent of the axioms of algebra, while giving context to practice.

Regarding the contents of this book, the first chapter deals with material that is required to understand the axioms of algebra (and much of modern, higher mathematics as well). Chapter 2 deals with the axioms relating to how numbers compare to one another, while Chapters 3 & 4 deal with the axioms that are more characteristically specific to algebra. Chapters 5 & 6 begin to take us beyond elementary algebra.

There are numerous practice problems at the end of each chapter. Some are designed to reinforce your understanding of the axioms themselves, while others are more traditional, designed to increase your mechanical proficiency, intuition and retention—retention comes from practice.

Conveniently, the order in which the axioms are presented in this book is aligned with the order in which they are traditionally presented; thus, finding more problems to accompany the axioms (or vice versa) should be relatively easy. Once mastered, please feel free to use a calculator or computer at your convenience.

It is important to point out, however, that the axioms of algebra, indeed the axioms of any mathematical subject, are assumed to be all present, ready and available from the get-go; but, of course, they can only be presented linearly. Although I have tried throughout the course of this book to examine and utilize each axiom to its fullest before presenting the next, and to only use previously examined axioms in proofs, it was occasionally necessary to call upon a subsequent, obliging axiom or two to explain the axioms at hand more completely. However, I trust that my audience has some familiarity with algebra and are, therefore, able to persevere. However, reference is made to the obliging axioms wherever applicable.

I have also, somewhat early on, for the sake of clarity and flow, introduced two assumptions in the book, the proofs of which cannot be exhibited until the section on the Distributive Axiom (Chapter 4.6) has been taken up. However, they should be very familiar to you: (1) every number multiplied by zero is zero; and (2) every number multiplied by negative one is the negative of the number.

You may safely overlook some of the sections in this book (although I urge you to read them all) depending on your confidence level of the subject matter contained therein. Thus, you may choose to skip Chapter 1, Section 2 on logic if you believe that you are equipped with most of the logical tools required to understand the axioms of algebra. However, logic does have its subtleties, so be warned. If you have a decent knowledge and understanding of the concept of sets, then you may also wish to skip Chapter 1, Section 3 as well.

Answers to the practice problems are given at the end of the book. Throughout the book, you may also discover stars, ☆. They indicate deeper, theoretical problems whose answers can be found in Appendix A. Appendix C lists all the axioms, definitions and theorems contained in this book. It would do you well to reference it often while reading.

~

CHAPTER 1
PRELIMINARIES

1.1 NUMBERS

It is remarkable that so many of us have completed high school algebra without ever truly knowing what numbers are. We have gone about our studies adding and multiplying them, squaring and cubing them, and pointing them out on a number line, without ever fully understanding the objects of our study. To be sure, this was not an oversight of our teachers or parents. Rather, it is because of the diaphanous nature of numbers. One might as well ask a child to describe the color of wind, while commenting to the child, "*I-can't-see-it* is not a color."*

Even as we have matured in our education of mathematics, the reality of numbers, if there is such a thing, continues to evade our senses, although we often believe otherwise. As a reminder of its delusory nature, however, I ask that you indulge me for a moment and hold up the index finger from each hand and then add them. If you are confused about how, exactly, to accomplish this request, it is because only numbers can be added, and fingers, need I remind you, are not numbers. That being the case, we may wonder whether numbers exist at all.

However, if they do exist, then they can only exist in the ethereal realm of Plato's perfect circle, far beyond the jagged edges of reality, because in truth there is nothing on earth or in the cosmos that we can point to that possesses the characteristics of *number*. Nevertheless, whatever wispy existence and structure these ghostly apparitions may have, they appear to have tremendous influence on life and the Universe at large.†

But I digress from my point, which is that numbers, like perfect circles, are ideas. All that we have, or ever will have, are the words for numbers, and the ability to communicate our ideas about numbers through language—as esoteric as it may often be—and reason. This is an important fact about numbers, and about the algebraic language we will be using to not only define them, but to discover their secrets, and it requires some deeper consideration.

Consider the symbols 5 and 3, for example, and the two operations of arithmetic: addition and multiplication ($+$ and $\times$ are called operators). It was no small feat that somewhere around the time of kindergarten we learned that

---

* Numbers are a child's first introduction to the world of abstract mathematical thinking.

† For example, does life only emerge on the border between chaos and order? Why is the Golden Ratio (Chapter 5) and the Fibonacci sequence (1,1,2,3,5,8,13,...) so prevalent in nature? (The Golden Ratio is even found in the structure of DNA's double helix.) Could there be a single equation (the holy-grail equation) that doesn't just describe the cosmos, but also governs it?

the symbol 5 is the word we use to represent the idea of the number five and that the symbol 3 is the word we use to represent the idea of the number three.

In time, we were also able to make sense of the sequence of symbols $5+3$ and 5×3, and having thoroughly learned the language behind the symbols, we were able to perform the indicated operations and recognize, through counting and memorized facts, that these symbols were just another way of writing the *ideas* of eight and fifteen. In fact, we have learned our lessons so well that whenever we see the symbol for a number, 8, 15, 77, etc., we automatically call the symbol itself a number.

Although symbols (numerals) are not, linguistically speaking, numbers any more than the word *cat* is the adorable, purring, household creature, practically speaking there is no reason why we shouldn't call them numbers, and for reasons of logical necessity, as we shall learn, we must consider them numbers, because the axioms of algebra apply as much to these symbolic representations of the numbers as they do to the idea of the numbers themselves and from a mathematician's point of view, perhaps, even more so.

Interestingly, despite the infinitude of them, numbers also suffer a type of desultory existence. Like picking rummage at a church tool-sale, people pick and choose numbers for whatever multitude of reasons they need them with little regard to how they are organized. They choose numbers to calculate the food stores necessary to make it through winter, to impose tax on a worker's labor, or to estimate equivalent value when negotiating a fair-trade agreement, but rarely do they ever consider the structure from which these numbers come.

In the mind of a (theoretical) mathematician, however, this is totally unacceptable. If any truth about numbers is to be ascertained, then numbers need to be reined in and structured, and above all they need to be defined. More than anything else, the axioms of algebra, taken in their entirety, not only structure numbers, but also tell us precisely what numbers are, and what they are not.

Fortunately, we do not have to read through the entire set of axioms contained in this book to comment on the two types of "numbers" used in algebra. They are (1) the familiar numerals of arithmetic, and (2) they are the letters of our modern Western alphabet, and the Greek alphabet as well, which on occasion I will call alpha-numbers to distinguish them from numeral-numbers. Thus, numbers like a, b, c, x, y, z, α, β, θ, etc. are alpha-numbers, while numeral-numbers are of the sort 0, 1, $\sqrt{2}$, 7, π, -213, $7/8$, 32.537, etc.

Alpha-numbers, however, would be better characterized as pseudo-numbers because, unlike numeral-numbers, they only have the *presumption* of being numbers. They are called into service only for the sake of some logical argument. For example, if x were a number, and strictly speaking it is not (it is a *logical variable* – a symbol used to replace an idea, a number, or string of words), but *if* it were a number and, say, $2x+1=15$, then we could ask the question, "what numeral-number does x represent?"

To discover the number, we could of course guess, and that is a perfectly acceptable and natural thing to do. Much of mathematics is done by guesswork and laborious strings of calculations in the hope of finding some revealing pattern, but guessing is a needlessly hit or miss proposition when it comes to algebra, which is well grounded in theory and practice and has long stood the test of time.

Alternatively, we could assemble all the characteristics or properties that all numeral-numbers have in common and use them, in a sense, to "reverse engineer" the equation and thereby logically deduce, rather than guess, the unknown numerical-number. However, for that to work the alpha-number, x, must be endowed with all the same common properties that the numeral-numbers have, and furthermore, each numeral-number must have unique characteristics of its own that distinguishes it from every other numeral-number.

The first part of this requirement is easily facilitated with the word *if*. Thus, *if* x were a number, then $x+2=2+x$ and $x+0=x$, because these are common properties of all numbers, and *if* x were a number, then it would have these properties as well. We make assumptive sentences like this all the time without giving it a moment's notice. For example, if I were a bird, then I could fly above the trees and look down upon the earth.

The second part comes to us by way of history. Each numeral cannot represent any other idea or number than that to which it has been historically assigned. Thus, each numeral-number is unique. For example, the number 7 has the unique property such that it divides the number 21 into 3 equal parts—no other number possesses this property. Likewise, the number 139 has the unique property that when we add 3 to it, we get the number 142; and so it is with all the numeral-numbers. If this were not the case, then there would be no way of determining the *unique* numeral-number that x represents in the equation $2x+1=15$ —and our world would be fantastically different.

This reveals an important distinction between the two types of numbers: alpha-numbers, unlike numeral-numbers do not have any of the unique characteristics that define each numeral-number. However, it is precisely *that* characteristic of the alpha-numbers that gives them the flexibility to assimilate the unique characteristics of a specific numeral-number whenever they are either logically discovered to be that number or are deliberately assigned to be that number. Thus, if x is discovered or assigned to be 7, then x also logically divides 21 into 3 equal parts because that is a unique characteristic of the number 7, which the symbol x now possess (and symbols are symbols, whether they look like "x" or "7"); or if x is discovered to be the number 139, then we know that $x+3=142$, because that is a unique characteristic of 139.

In summary, by permitting alphabet letters to act as surrogate numbers, we have tacitly conferred upon them all of the properties that the numeral-numbers have in common, like the commutative, associative and distributive properties to name a few. In doing so, we have not only provided a foundation for logically discovering the relations between the alpha-numbers and the numeral-numbers in any given equation, like $2x+1=15$, but we have also provided a method for discovering the hidden and unique numerical-number that an alpha-number often represents.

These common properties were abstracted from numbers through the many millennia that human beings have been using numbers, and at least one of them, the commutative property of addition, was no doubt perceived, if only subconsciously, by every human that could count, even, I would argue, those humans dating back to Man's most primitive origins. However, the larger part of the history of numbers has concerned itself with the unique properties of each number and not their common properties. It is only within the last three centuries that mathematicians have had a need to consider the common properties more prominently.

Furthermore, although the axiomatic method has been around since, at least (arguably), the time of the Egyptians, it too began to take deeper root in mathematics around the nineteenth century, whereupon the common properties of numbers became the axioms of algebra.

The common properties are considered the basic tenets of algebra and are the mathematician's general definition of the word *number*. Thus, any collection of mathematical objects that fails to abide by or conform to *every* axiom will not be considered a viable collection of numbers.

As you continue your reading, you should try to keep in mind that the word *numbers* always needs to be viewed on two levels. On one level, it has an interpretation—it has meaning—it is about our collective idea of numbers in their relation to our commonplace understandings of equality, inequality and the operations of addition and multiplication—it is about all the things we have always known numbers to be about. On another level, however, the word has no meaning at all. It represents a collection of abstract mathematical objects—symbols—to be manipulated according to the rules of logic with special attention to, at the very least, the four symbols $=$, $<$, $+$, and $\times$ (or $\bullet$) .

~

1.2 Logic

The benefit from studying logic is that it provides the tools necessary for good reasoning skills, and no other subject is so wholly dependent on proper reasoning than mathematics. Unlike physics, chemistry, biology, psychology, sociology and so many other sciences, mathematics is bereft of the luxury of experimentation that would allow a mathematician to determine the correctness of his ideas, theories and beliefs. For that reason, he must start from a position of what he believes to be true in his heart of hearts and use logic to deduce everything else.

However, the subject of logic is much too vast for anything but an abridged overview of just those aspects that will help you to understand the mathematical proofs contained in this book. To that end, we will take a brief look at Aristotle's syllogistic logic, modern propositional logic, the four conditional statements of logic, the use of the logical conjunction *and,* the disjunction *or,* and *proof by contradiction*. Most of us have a natural sense of logic, so hopefully that will be good enough to see you through to the end.

The use of logic for discovering truths in mathematics is at least as old at the Ancient Egyptians and the Ancient Greeks. The Greeks were especially aware of the utility of logic as exemplified in Pythagoras's Theorem, Archimedes' writings in geometry, Euclid's *Elements of Geometry,* and most important, certainly for our purposes here, Aristotle's codification of the rules of logic.

For his part, Aristotle's most notable contribution to the field of logic is the *three-line syllogism*. However, because not every logical argument can be written as a three-line syllogism, much of Aristotle's syllogistic logic has been replaced with propositional logic, the differences between which we will not elaborate on here. However, because every syllogistic argument is equivalent to a propositional argument, this form of reasoning is, for the most part, as valid today as it was when Aristotle first illuminated it. For example, the following three-line syllogism consists of two declarative sentences, called *propositions,* followed by a conclusion:

turtles are reptiles (major proposition/premise),
reptiles are cold blooded animals (minor proposition/premise), ▪1.2.1
therefore, turtles are cold blooded animals (conclusion).

It is a quirk of language that by reversing the order of the words in either of the first two statements in 1.2.1 the meaning is completely changed. Thus, *reptiles are turtles* seems wrong, while the other way around seems right. The reason is because we naturally tend to put things in categories or sets*.

Constructing *Venn diagrams,* named after John Venn who formalized the utility of these diagrams to show the logical relations among the various categories (otherwise known as terms) in a logical argument, can often aid your understanding whenever you come across a more difficult logical argument.

At the top of the opposing page is a Venn diagram depicting the nested categories that the terms, turtles, reptiles, and cold-blooded animals from syllogism 1.2.1 belong to.

* The idea of logical categories comes down to us from Aristotle.

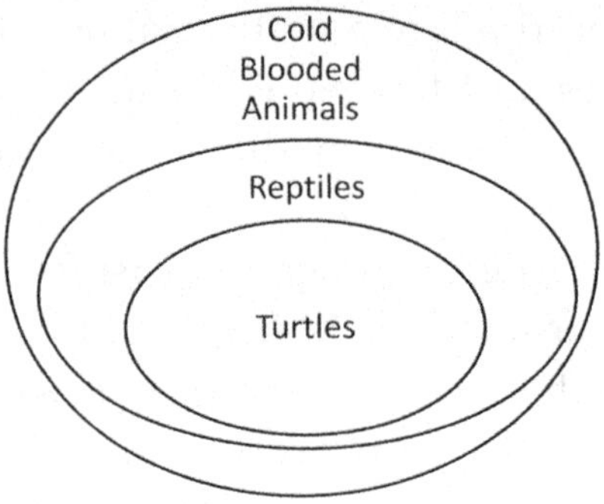

From the diagram, it is easy to see that turtles are a subclass of reptiles, and that reptiles are a subclass of cold blooded animals, and that therefore, turtles are a subclass of cold blooded animals.

Although it seems trivial to mention that there would be nothing to deduce if it were not for the first two statements in a three-line syllogism, it is an essential characteristic of the axiomatic method; deduction always requires one or two initial propositions that we assume to be true—something to prime the deductive pump with, so to speak.

These propositions are "interlaced" in such a way that, when it comes to syllogisms at any rate, we can make the deductive leap and arrive at a conclusion. Thus, if we care to, we can deduce that the sun is hot—provided we have a good opening line or two:

if it melts chocolate, then it is hot,
the sun melts chocolate,
therefore, the sun is hot.

A more relevant syllogism for our purposes would be the following:

if a and b are numbers, then $a+b=b+a$,
1 *and* 0 *are numbers,*
therefore, $1+0=0+1$.

Why we should accept the syllogism, as well as any other rule of logic, as a valid form of reasoning, especially given the mercurial nature of humankind, is a matter for philosophers, neurologists, linguists and psychologists. Perhaps it has to do with our deep-seated belief that for every effect in the universe there is a cause—Newtonian physics at its finest. Regardless, it is within the range of human experience and thought to intuitively understand the logic behind syllogisms and agree with their pronouncements.

Aristotle's syllogisms, although marvelous and insightful on their own, especially since they were written well over two millennia ago, nevertheless, lack a certain sophistication that is essential for modern science and mathematics. In particular, Aristotle makes only slight mention of the concept of *induction*, being more concerned, as he is, with matters of *deduction*.

Deduction, as in the syllogisms supplied previously, begins with a few premises from which the stated conclusion must necessarily follow. Induction, however, is the act of attributing a declarative statement to an entire collection of objects based on many or even only a few evidentiary observations of that object—it is an act of faith. For example, the following argument is an inductive argument:

Today I saw a green, toothless turtle. Yesterday I saw a slimy, toothless turtle. The day before that I saw a tailless, toothless turtle, and the day before that I saw a dead, toothless turtle. Therefore, all turtles are toothless.

Whether this argument can be supported by evolutionary or other theories about turtles is irrelevant, its truth ultimately depends on whether there exists a hitherto-unseen turtle with teeth—a *counter example.*

Although most modern college-level textbooks cite the principle of induction as an axiom of algebra and specify it much more precisely and cleverly than our paltry example given above can convey, we will excuse ourselves from citing it in this book, noting that it is not essential for our purposes. We will rely on good common sense and a smattering of mathematical intuition to infer that some things are true for all occurrences, when having only encountered a few.

However, induction is not the only departure from Aristotelian logic that modern logic makes. Aristotelian logic would never have had one of the premises as the conclusion. Hence, the argument,

turtles are reptiles (major premise),
reptiles are cold blooded animals (minor premise),
therefore, turtles are reptiles (conclusion),

would have never appeared in any writings about logic prior to the middle of the nineteenth century. Yet, if we wish to cover *all* the bases in logic, then we should include a deduction of this sort into our catalogue of valid forms of deductive reasoning; as well as when the argument has only one premise:

turtles are reptiles,
therefore, turtles are reptiles.

This statement is patently valid and true. We should also include those statements where neither the premises nor the conclusion is true, but where the argument, itself, still manifests a sense of validity; for example,

turtles are birds,
birds are hatched on the moon,
therefore, turtles are hatched on the moon. ▪1.2.2

It should be clear that the truth or falsehood of the premises and conclusion in this argument has no effect on the *validity* of the argument, itself. That is, regardless of the actual truth of each of the above statements, the argument is still valid because *if* the two premises *were* true, and we always start with the assertion that they are, then we would be compelled to conclude that turtles are hatched on the moon.

Logicians define a distinction between *valid* arguments that are true and those that are false. The former are considered *valid and sound*, while the latter are considered *valid and unsound*.

∘ DEF. 1.2.1: SOUND AND UNSOUND ARGUMENTS

A deductive argument is sound if and only if it is both valid and all of its premises are factually true, otherwise the argument is unsound.*

If we take this definition apart, you will notice that it says nothing about the conclusion of an argument. That's because if the premises are true and the argument is valid, then the conclusion must be true.

Based on this definition we may conclude the following statements.

1. All sound arguments are valid.
2. All invalid arguments are unsound.
3. Some valid arguments are unsound.
4. Some unsound arguments are valid.

Some of you may have also noticed the futility in all of this—we are relying on your innate understanding of logic to explain and describe for you the concepts of logic. Unfortunately, there is no easy way to get around this merry-go-round, but that did not discourage some great logical minds from trying. The story, although slightly tangential to the subject at hand, lends direction and depth to the subject of logic, so we shall take a moment to consider it.

Around 1911, Alfred North Whitehead and Bertrand Russell had produced three massive volumes (entirely unreadable except by zombies) describing and reducing all the symbols and rules of mathematics to the symbols and rules of logic (modern versions of symbolic logic have long since moved on). Their work, titled *Principia Mathematica* (PM), contained page after page of symbols, a section of which might look something like the image created below,

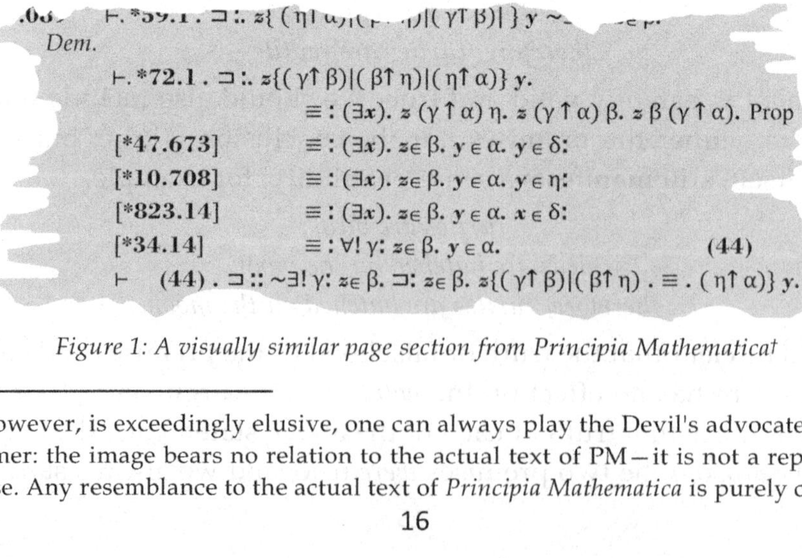
Dem.

⊢.*72.1 . ⊃ :. z{(γ↑β)|(β↑η)|(η↑α)} y.

≡ : (∃x). z (γ↑α) η. z (γ↑α) β. z β (γ↑α). Prop

[*47.673] ≡ : (∃x). z ε β. y ε α. y ε δ:

[*10.708] ≡ : (∃x). z ε β. y ε α. y ε η:

[*823.14] ≡ : (∃x). z ε β. y ε α. x ε δ:

[*34.14] ≡ : ∀! γ: z ε β. y ε α. (44)

⊢ (44) . ⊃ :: ~∃! γ: z ε β. ⊃: z ε β. z{(γ↑β)|(β↑η) . ≡ . (η↑α)} y.

Figure 1: A visually similar page section from Principia Mathematica†

* Truth, however, is exceedingly elusive, one can always play the Devil's advocate.

† (Disclaimer: the image bears no relation to the actual text of PM—it is not a reproduction or paraphrase. Any resemblance to the actual text of *Principia Mathematica* is purely coincidental.)

Unfortunately, their nobly inspired and masterfully crafted, ten-year project was doomed from the moment they set pen to paper because (although it was unknown at the time) hidden within any description of *every* sufficiently complex set of symbols and rules lurks an inherent and inexorable circularity. It turned out, as it were, that the symbols and rules of logic were no less mathematical than the symbols and rules of mathematics itself, which Kurt Gödel conclusively demonstrated in his magnum opus, *On Formally Undecidable Propositions of 'Principia Mathematica' and Related Systems.* To say it bluntly, the two logicians were using the rules of logic to deduce the rules of logic.

The moral of the story is that *every* consistent (i.e., non-contradictory) and robust system of axioms, like those in algebra, is incapable of answering every question regarding that system. In other words, there are some "well-formed" statements or conjectures* that are impossible to either prove or disprove using the axioms of algebra—such was Gödel's proof. Unfortunately, I cannot present you with an example of such a statement because it would require *proving* that the statement is neither provable nor disprovable. Regardless, we have no choice but to carry on and prove what we can.

When zombies go to the library.

When it comes to proof, however (to continue), Aristotle's theory of logic, by itself, does not contain all the requisite tools for mastering it. We also need to understand the statements or propositions that make up his syllogisms, which can be anyone of the four conditional statements called the converse, the inverse, the contrapositive and the original statement.

A *conditional statement* or *proposition* is any statement that has an "if ..., then ..." form to it. The first part of a conditional statement, the "if" part, is called

* Statements that are not only syntactically correct algebraic statements (i.e., there are no misplaced addition or multiplication operators, equality symbols, parentheses, etc.), but also semantically correct algebraic statements (i.e., they also have meaning.).

the *antecedent* or the *premise;* the second part, the "then" part, is called the *consequent* or the *conclusion.* Thus, the statement,

if lightning strikes the ground, then thunder claps, ▪1.2.3

is a conditional statement for which the phrase *lightning strikes the ground* is the antecedent or premise, and *thunder claps* is the consequent or conclusion.

Language, and "Devil's advocates", will inevitably get in the way when trying to make sense of logical statements, consequently, assumptions run rampant. In the above statement, for example, we are assuming that *all* occurrences of lightning make thunderclaps, and that, unlike the if-a-tree-falls-in-the-forest scenario, whenever lightning does strike, it not only claps, but *everyone* can *always* see it and hear it too.

If we care to, we could make statement 1.2.3 into a three-line syllogism:

all lightning strikes produce thunderclaps,
lightning strikes the ground,
therefore, thunder claps.

An *original* or *initial* conditional statement is any conditional statement that appears first in a discussion. Since 1.2.3 was the first statement to appear in our discussion, we will consider it the original conditional statement, and, in this case, it is true.

However, the *converse* of this statement is not true. The converse states

if thunder claps, then lightning strikes the ground. ▪1.2.4

In fact, of the many times a thunderclap occurs, only rarely does lightning hit the ground. Thus, usually, if the original statement is true, the converse is not true, but there are numerous exceptions.

The *inverse* of the original statement is also false. It states

if no lightning strikes the ground, then no thunder claps. ▪1.2.5

It is possible for no bolt of lightning to strike the ground (a bolt could strike from cloud to cloud for example) for which there would still be a thunderclap (obviously). Hence, the inverse of statement 1.2.3 is false. We could say that lightning striking the ground is not a *necessary* condition for a thunderclap to occur because there are other conditions under which a thunderclap can occur; however, lightning striking the ground is a *sufficient* condition for a thunderclap to occur.

The truth or falsehood of the inverse of a statement is directly related to the truth or falsehood of the converse and, accordingly, may or may not be true independently of the truth or falsehood of the original statement.

However, there is one thing that we can say with absolute certainty based upon the original statement 1.2.3:

if no thunder claps, then no lightning strikes the ground. ▪1.2.6

The reason is because if lightning had struck the ground, then by our original statement thunder would have clapped. However, because no thunderclap occurred, lightning could not have possibly struck the ground (otherwise thunder would have clapped. *¡Ay, caramba!*). This is called the *contrapositive* of the original condition and it is directly true (or false) if the original statement is true (or false). It also lends merit to proof-by-contradiction, which we will discuss momentarily.

Due to the peculiarities of the English language we are often forced to change the tense of a statement to indicate the sequence in time between the cause of an event and its effect. This can often lead to confusion when examining the validity of an argument or a statement. For this reason, and because of the inherent assumptions we inevitably make when using language, it is a good idea whenever you are having trouble understanding a proof to replace the antecedent and the consequent of any conditional statement with logical variables and write instead, something like, "*if p, then q*", thereby avoiding the ambiguities of language. In our example p would represent the phrase "*lightning strikes the ground*" and q would represent the phrase "*thunder claps*".

To obtain the negation or opposite of these phrases we use the tilde symbol, "$\sim$", which means *not*. Thus $\sim p$ (not-p) means the negation of, or the opposite of "*lightning strikes the ground*", and $\sim q$ (not-q) means the negation or opposite of "*thunder claps*". Furthermore, the negation of a negation is just the thing itself. Thus, $\sim(\sim p)$ is just p. If I were to comment any further, I would only be inviting the many eccentricities and ambiguities of language into the discussion, which is the very thing I am trying to avoid in the first place. Often it is best to just go with your first impressions and let your intuition prevail.

◦ DEF. 1.2.2: CONDITIONAL STATEMENTS—ORIGINAL, INVERSE, CONVERSE, AND CONTRAPOSITIVE

A conditional statement is any sentence or declarative statement that relates a cause to the effect of that cause. The following are conditional statements,

1. Original: *if p, then q* (or "*p implies q*", or "*q, if p*")
2. Converse: *if q, then p* (or "*q implies p*", or "*p, if q*")
3. Inverse: *if ~p, then ~q* (or "*~p implies ~q*", or "*~q, if ~p*")
4. Contrapositive: *if ~q, then ~p* (or "*~q implies ~p*", or "*~p, if ~q*")

Any of the statements, 1.2.3, 1.2.4, 1.2.5 or 1.2.6, could have been chosen as the original statement, upon which all of the other statements would have to be relabeled accordingly. Thus, if 1.2.4 were the original conditional statement, then 1.2.3 would be its converse; 1.2.6 would be its inverse; and 1.2.5 would be its contrapositive.

Furthermore, the *truth values* of statements 1 and 4 are equal. In other words, the original statement is true if and only if the contrapositive is true. In the same manner, the truth values of statements 2 and 3 are equal.

Before moving on, I should briefly explain the meaning of the phrase *if and only if* (often abbreviated as *iff*) as used in the preceding paragraph (as well as when used in Def. 1.2.1). We will proceed by example. If the statement,

the bee flies if and only if the bee has wings,

is true, then the following two statements are also true:

if the bee flies, then the bee has wings (original);
if the bee has wings, then the bee flies (converse).*

A true if-and-only-if statement is, accordingly, a composition of a true original and a true converse of the statement. Because the contrapositive is always true whenever the original is true, and because the inverse is always true whenever the converse is true, it must be the case that all four types of conditional statements are true whenever an if-and-only-if statement is true. Consequently, the proof of an if-and-only-if statement always requires proving both the original statement and its converse. (*Definitions rarely require proof.*†)

Conditional statements may also include the word *and* or *or* (or both) in various combinations, consider the following original statement:

If you read PM‡ <u>and</u> you eat brains, then you are a zombie. ▪1.2.7

It should be clear from the language used in this statement that it is not enough to just read PM, or to just eat brains to be a zombie, one must do both to be considered a zombie.

Equivalently, we can express a related thought with the contrapositive:

if you are not a zombie, then you do not read PM <u>or</u> you do not eat brains.

Notice that we had to replace the conjunction *and* of the antecedent from the original statement with the disjunction *or* in the consequent of the contrapositive statement. In other words, *the negation of a conjunction is the same as the disjunction of the negations.* In symbols, we could express this relation as

$\sim(p \text{ and } k) \Leftrightarrow \sim p \text{ or } \sim k \text{ (or both)}$§ ▪1.2.8

For example, not having both a potato *and* a carrot is logically the same as not having a potato (or not having a carrot) because if you have a carrot and no potato (or a potato and no carrot), then you obviously do not have *both* a potato and a carrot. You have essentially negated the conjunction *and* whenever you don't have one or the other. Of course, if you have neither a potato nor a carrot, then it goes without saying that you do not have both.

---

* This example illustrates the complications of language. One may argue that a dead bee, which has wings, certainly cannot fly. However, as mentioned earlier, we are not playing the Devil's advocate.

† Occasionally, a definition may call an object into existence, proof of existence is then sometimes required.

‡ PM: *Principia Mathematica* by Whitehead and Russell.

§ The symbol "⇔" is the symbolic shorthand for "is equivalent to" or for "if and only if".

The word *or*, as used in logic and mathematics, (almost) always includes the possibility of both. Thus, if we agree that if it's raining or snowing, then you will get wet, then you will certainly get wet if it's doing both.

I leave it to you to write out the converse and the inverse to the original statement 1.2.7. (☆1)

Now suppose, we have the following statement:

if you read PM <u>or</u> you eat brains, then you are a zombie. ▪1.2.9

This statement is logically different than statement 1.2.7 because both conditions are not required to be a zombie—it is enough to just be a brain eater (or a PM reader) to be a zombie. However, for this statement the contrapositive is

if you are not a zombie, then you do not read PM <u>and</u> you do not eat brains.

Here, because we have negated both the antecedent and consequent of the original statement (1.2.9), which is what the contrapositive does in addition to exchanging the antecedent for the consequent and vice versa, we have replaced the negation of a disjunction with the conjunction of the negations. In symbols, we would write

$$\sim(p \text{ or } k) \Leftrightarrow \sim p \text{ and } \sim k. \qquad ▪1.2.10$$

Note that $\sim(p \text{ and } k)$ is logically *not* the same as $\sim p$ *and* $\sim k$. $\sim(p \text{ and } k)$ means that *not both p and k <u>may</u>* occur (not both may occur). Hence, if only one of them, say only *p*, occurs, then certainly *not both* occur. Contrastingly, "$\sim p$ *and* $\sim k$" means that *both* $\sim p$ *and* $\sim k$ *<u>must</u>* occur (both negations must occur).

Knowing the equivalent statements for the negation of a conjunction and the negation of a disjunction, statements 1.2.8 and 1.2.10 respectively, we can simplify more complex statements involving both *and* and *or*. Thus,

$$\sim((p \text{ or } n) \text{ and } k) \Leftrightarrow \sim(p \text{ or } n) \text{ or } \sim k \Leftrightarrow (\sim p \text{ and } \sim n) \text{ or } \sim k.$$

The parentheses are necessary to distinguish between what the conjunction, *and*, pertains to and what the disjunction, *or*, pertains to.

We can gather up the conditional conjunction and the conditional disjunction statements along with their counterparts in a convenient table.

STATEMENTS	AND	OR
◇ Initial	*If p and k, then q*	*If p or k, then q*
◆ Converse	*If q, then p and k*	*If q, then p or k*
◆ Inverse	*If ~p or ~k, then ~q*	*If ~p and ~k, then ~q*
◇ Contrapositive	*If ~q, then ~p or ~k*	*If ~q, then ~p and ~k*

Table 1 – Conditional-and, and conditional-or statements

Our last topic of discussion for this section is the logic behind proof-by-contradiction. It is the mathematical equivalent of the process of elimination. In the words of Sherlock Holmes, "once you eliminate the impossible, whatever remains, however improbable, must be the truth".* As a quick example, suppose we wanted to show that $0<1$. Because there are only three possible ways in which the number 0 can compare to the number 1, either $0>1$, $0=1$ or $0<1$, and only one of them is the correct comparison, it follows that if we can eliminate both $0>1$ and $0=1$, then the correct comparison must be what remains, namely $0<1$.

For a more detailed and general discussion however, suppose we wanted to prove the original statement: *if p, then q*. Now, whatever p and q represent is not particularly important, what is important is that we accept p as true because if we don't, then there would be no point in commencing the argument in the first place†, not to mention that we couldn't be sure of the truth of q. Hence, we always start from a premise that we assume, or assert is true. For example, suppose we had the statement: *if turtles had teeth, then they could chew.* Of course, we all know that turtles do not have teeth, however, if we do not accept the premise as true, then there is no argument.

Now, we could try to prove the truth of q directly from p by proceeding in a logically valid, linear, step-by-step fashion; or, because the truth value of the contrapositive is identical to the original, we could try to directly prove the contrapositive, which is *if ~q, then ~p*. However, direct proofs are often difficult to achieve. So alternatively, we could combine the two statements and see if we run into a contradiction because if *p implies q* is true, then *p <u>and</u> q* must be true, while *p and ~q* must be false‡. In other words, *p and ~q* must give rise to a contradiction. Analogously then, *if the bee flies, then it has wings* is true, then *the bee flies and it has wings* must also be true, while *the bee flies and it has no wings* must be false. Thus, by proving the latter statement false, which is done by contradiction, we imply the truth of the former.

For example, suppose we have prior knowledge—proof if you like—that zero times any number is zero§ (i.e., if $a=0$ or $b=0$ [or both], then $ab=0$**), and we wish to prove

$$\textit{If } a\neq 0 \textit{ and } b\neq 0, \textit{ then } ab\neq 0. \qquad \blacksquare 1.2.11$$

This statement would be difficult to prove directly because, even though experience has taught us that whenever we multiply any two *non-zero* numbers we get a non-zero number, we haven't tried *every* pair of such numbers. The exception—the counter example—although obvious to an intuitionalist that it

---

* *The Hound of the Baskervilles* – Sir Arthur Conan Doyle

† Even if p were a false statement, we would accept the false statement as true.

‡ This is known as the law of the excluded middle: either an object possesses a property, or it does not—there is no middle ground. There is philosophical debate over this law and its implications, which has given rise to a branch of mathematics called Fuzzy Logic.

§ This needs to be proved (Theorem 4.6.1), which we will later, but for now we will accept it as common knowledge.

** This statement is the inverse of 1.2.11.

doesn't exist, could, in the mind of logician, still be lurking in the densely-packed, infinite forest of untried numbers. The same dilemma makes it difficult to directly prove the contrapositive of statement 1.2.11:

$$\textit{if } ab=0, \textit{ then } a=0 \textit{ or } b=0 \textit{ (or both).} \qquad \blacksquare 1.2.12$$

In either case, our knowledge that zero times any number equals zero is of little immediate use because it happens to be the inverse of our original statement (i.e., 1.2.11), and the truth value of the inverse of a statement is not related to the truth value of the statement. As a last resort, we may consider trying to prove 1.2.11 by contradiction.

Hence, we proceed on the assumption that the premise from 1.2.11, $a \neq 0$ and $b \neq 0$, *and* the negation of its conclusion, which would be $ab=0$, are both (i.e., all three equations) true, which has the effect of creating the aforementioned, logically impossible situation. As you can see from statements 1.2.11 and 1.2.12, $a \neq 0$ and $b \neq 0$ and $ab=0$ cannot all be true if these two statements, whose truth values are directly connected, are true. Nevertheless, we continue logically from this supposition as follows: because $a \neq 0$, we are allowed to write $1/a$ * (only with the number 0 are we not allowed to do this and a is not 0). Furthermore, because $ab=0$, we have $(1/a) \cdot ab = (1/a) \cdot 0$ † (we are allowed to multiply both sides of an equality by the same number without changing the truth value of that equation). On the left hand side of this equality, we have $(1/a) \cdot a \cdot b = [(1/a) \cdot a] \cdot b = [1] \cdot b = b$, that is, we have b on the left; while on the right, after utilizing our knowledge that zero times any number equals zero, we have 0. Putting it all together we must conclude that $b=0$, but this contradicts our premise that $b \neq 0$. Had we started the proof a little differently, that is with $b \neq 0$ (which also has a multiplicative inverse), we would have arrived at the contradiction‡ that $a=0$. Consequently, it must be the case that our assumption that $ab=0$ was a false assumption, ergo $ab \neq 0$.§ Thusly, we have proved by contradiction statement 1.2.11, which is what we set out to prove.

~

1.3 SETS

Sets are foundational and indispensable to almost every branch of mathematics, including algebra. They provide us with a language for communicating our thoughts and ideas about classes or collections of objects along with the relations they may exhibit among themselves. Some knowledge of sets is therefore unavoidable, and the more knowledge the better. Set theory is also

* Here we have made use of the assumption that $a \neq 0$.
† Here we have made use of the assumption that $ab=0$.
‡ because of the premise's symmetry.
§ We will prove statement 1.2.11 more formally when we get to Chapter 4.

a branch of mathematical logic, and rightly so because many of the ideas from logic, like the categories and subcategories of Aristotle's syllogisms, parallel the ideas of sets and subsets.

Set theory was developed by the mathematicians Georg Cantor, Giuseppe Peano and Richard Dedekind, and a few others, while the formal (symbolic) axioms governing set theory were developed by Ernst Zermelo and later improved upon by Abraham Fraenkel. It "settled"* many of the arguments that mathematicians had regarding the concept of mathematical infinity and some paradoxical problems involving the language of mathematics. It finds its way into algebra in the form of number sets, number intervals, functions, the range and domain of functions, and solution sets.

The formal axiomatic rules that govern sets are a bit too abstruse for anything but a brief introduction to some of the concepts, operations and utility of sets. What this discourse lacks in depth, however, will, hopefully, be a gain in clarity.

We are all familiar with sets in one form or another: a collection of Barbie dolls, a set of box wrenches, or a set of silverware dining utensils to name a few. In the case of a silverware set, for example, the elements, otherwise known as the members, or items of the set are the various serving and carving utensils, dinner forks, knives and spoons, some of which are repeats depending upon the number of people that the set is designed to serve. Hence, a set that serves ten will have ten dinner knives, ten salad forks, etc.

However, even though there may be many repeats, every item in the set is unique in its appearance for one reason or another. Some items will have tarnished more than others, while others will have slight deformations and scratches. Some will also differ in appearance because of their differing functions. Hence, a fork looks different from a spoon precisely because it has a different function than a spoon.

Rather than look at the entire set of silverware pieces, however, we may be more interested in the set that contains only one representative from each of the functionally unique items. So, we go about removing all of the functionally equivalent repeats and keep only one of each, which is essentially a silverware serving set for one person. Our silverware set will then consist of the eating utensils like those shown in Figure 2, where each item in the set is both visually and functionally unique.

For simplicity's sake, rather than drawing the set or writing its long description every time we wish to talk about "The Set of Commonplace Silverware Serving Pieces", which is certainly a mouthful, we will just conveniently call it the set S.

* The "jury" is still deciding on the best set of axioms to do the job, but it is a matter of preference.

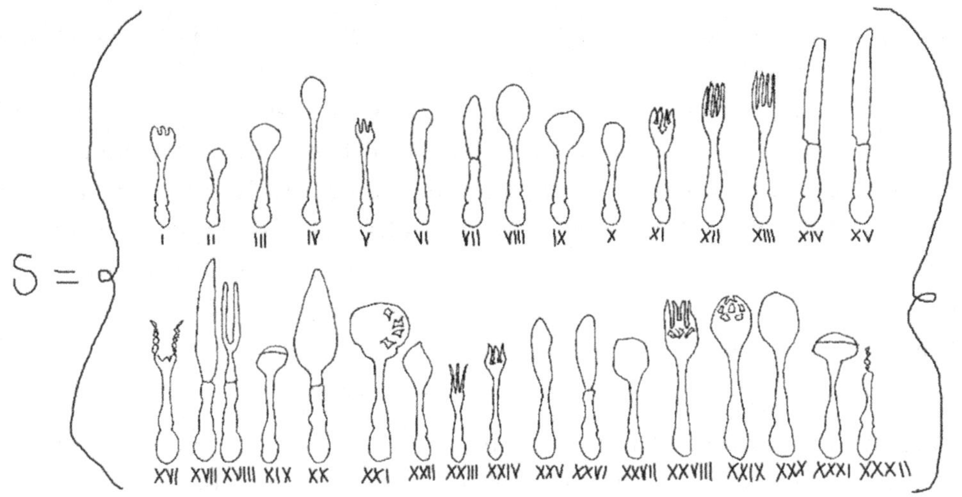

Figure 2: The Set, S, of Commonplace Silverware Serving Pieces (who knew?).
{I = ice cream fork; II = demitasse spoon; III = bouillon spoon; IV = ice tea spoon; V = cocktail fork: VI = flat handle butter spreader; VII = hollow handle butter spreader; VIII = desert spoon; IX = cream soup spoon; X = teaspoon; XI = salad fork; XII = regular fork; XIII = dinner fork; XIV = regular knife; XV = dinner knife; XVI = baked potato fork; XVII = carving knife; XVIII = carving fork; XIX = sauce ladle; XX = cake server; XXI = tomato server; XXII = jelly server; XXIII = lemon fork; XXIV = pickle fork; XXV = flat butter knife; XXVI = master butter knife; XXVII = sugar spoon; XXVIII = cold meat fork; XXIX = sieve silver spoon; XXX = serving spoon; XXXI = gravy ladle; XXXII = butter pick}

It is customary* to use a pair of curly braces, {,}, to denote the idea of a set. Thus, "{Commonplace Silverware Serving Pieces}" represents the set that consists of the commonplace silverware serving pieces; and because $S=${Commonplace Silverware Serving Pieces}, S is the set that consists of the commonplace silverware serving pieces.

To indicate that an item is an element of a set we use the curvy-E symbol, $\in$.† Thus we could write *cream soup spoon* $\in S$ to mean that the cream soup spoon is an element of the set S. Of course, we could also use the Roman numeral index assigned to each silverware piece to represent any of the elements of the set instead of the names of each piece, which would make our discussion about the elements and the set a bit more concise. Thus, we could write, $\mathrm{IX} \in S$ to say the same thing about the cream soup spoon being an element of the set S. We could also write $x \in S$ whenever we wish to talk about any piece of silverware, x, within the set S, but no piece in particular.

In doing this we have not only replaced all of the English with a few symbols, but we have made our discussion about sets more general and, unavoidably, more abstract. By *general,* I mean it is applicable to a wide range of uses

---

* The International Organization for Standardization (ISO) determines many of the conventions used in mathematics as well as the other sciences.

† The symbol " $\in$ " as used in the context of sets is a single character word. The English language has two single-character words: "I" and "a". Russian has ten. The language of mathematics has a gazillion. Some examples are, $7, x, \in, +, <, \neq, \mathbb{Z}, \surd, \Pi, \Rightarrow$; and many of these will have different meanings depending upon context.

and interpretations, mathematical or otherwise. By *abstract*, I mean it is a structure built of symbols.

Because the concepts involved in set theory are very general, they can be applied to numbers in much the same way that they can be applied to silverware, even better, because numbers seem to fall naturally into different classifications. There is the set of natural numbers, the set of even and odd numbers, the set of negative numbers, the set of prime numbers, of rational numbers, of numbers that end in 2, the set of numbers that are powers of 13, and so on.

However, consider the following set of numbers:

$$\{2,\ 81,\ 5,\ 34,\ 7,\ 30+4,\ 3,\ 9,\ 12,\ 3\times3,\ 17,\ 12+12+3+7,\ 3+2\}.$$

There are thirteen visually unique items or elements in the set, where each element is separated from the next by a comma. However, some of the elements in this set are functionally equivalent—they are repeats. The number thirty-four, for example, is present three times because each of the numbers 34, $30+4$ and $12+12+3+7$ function in the same way. Whenever we add any of these three numbers to another number, the number is increased by thirty-four and of course these three numbers function in the same way whenever they are subtracted from, multiplied with or divided by any other number. The same is true for the numbers nine and five in the set, each of which is functionally present twice. Hence, as with the silverware, we remove the repeats and keep exactly one of each. Denoting this set of numbers with T we have

$$T=\{2,\ 81,\ 5,\ 34,\ 7,\ 3,\ 9,\ 12,\ 17\}.$$

There is however a slight distinction to be made between a set of silverware pieces and a set of numbers: a set of numbers can be ordered* or at least it has a natural order. Thus, we can order the elements from least to greatest and write

$$T=\{2,\ 3,\ 5,\ 7,\ 9,\ 12,\ 17,\ 34,\ 81\}.$$

That is not to say that we cannot put the silverware pieces in some order; it is just that the order does not come naturally. We could, of course, order the silverware pieces according to length or weight or brilliance of finish. We could also order them according to the sequence in which they are used during a full course meal; by that account, the carving utensils might come first and the desert forks last. The point is that numbers lack any complicating physical attributes, and when we think of a set of numbers, it is natural to think of the set as ordered from lesser numbers to greater numbers, but that does not preclude the possibility of defining the order however we like or having no assigned order at all for that matter. Sets are defined by their elements and not how the

---

* However, sets that differ only in their order are still considered equal as we shall learn.

elements are ordered.

Most important, however, we need some formal way to describe the elements of a set. One way is to exhibit all of the elements (as we did with the silverware set) when the set contains a finite number of elements as, for example, in the set $E = \{2, 4, 6, 8, 10, 12, 14, 16\}$. Another way is to describe the defining characteristics of the elements of a set using a mix of the English language and mathematical symbols as in the set $E = \{$all $2x$, such that x is a natural number up to 8 inclusively$\}$.* We could also list just enough of the elements to clearly reveal some pattern among the elements, and then use ellipses to infer those elements not shown in the displayed set. For example, we can infer that the missing elements of $E = \{2, 4, 6, \cdots 14, 16\}$, are 8, 10, and 12. When the set contains an infinite number of elements, then the only thing that we cannot do is list or exhibit them. *Set builder notation* is another method for describing sets, but it is not needed for our discussion up to this point, so we will forgo it until the last chapter, at which point it will be most convenient to have.

Certainly, one of the first things we would like to do with sets, whether they are sets of numbers, silverware, kitchen curtains or whatever, is to compare them. A pair of sets can be either equal (When applied to sets, the word *equal* has a different interpretation then when applied to numbers.), disjoint, or one set can be a subset of the other (all of which will be defined presently). Otherwise they are just sets that have some elements in common and some not in common. Consider the following five sets of numbers in Table 2. (Spaces between the elements of the sets have been provided for your ease in making comparisons as was the order.)

$T = \{2, 3, 5, 7, 9, 12, 17, 34, 81\}$	9 elements
$Q = \{2, 5, 7, 17, 34, 81\}$	6 elements
$R = \{2, 5, 7, 17, 34, 81\}$	6 elements
$U = \{2, 7, 81\}$	3 elements
$V = \{1, 5, 15, 17, 49, 81\}$	6 elements
$W = \{8, 13, 19, 67, 95\}$	5 elements

Table 2: A Comparison of Six Sets

It should be obvious upon inspection of the above sets that sets Q and R are identical. Hence, we might consider a good definition for the equality between sets to be identicalness. Although it is true that when two sets are identical they are equal, it is not a mathematically workable definition because sets can have an infinite number of elements and, of course, we cannot inspect them all.

* Mathematicians define even, and odd numbers as follows: if x is a natural number or zero, then $2x$ is even and $2x+1$ is odd. (Thus, zero is an even number.)

A more workable definition, which allows for the possibility of being able to prove equality between sets even if they have an infinite number of elements, is as follows:

∘ DEF. 1.3.1: EQUALITY BETWEEN SETS

Two sets A and B are equal if and only if every element of A is an element of B and every element of B is an element of A. To be more mathematically precise we would write this as $A = B$, if and only if $x \in A$ implies $x \in B$ and $x \in B$ implies $x \in A$.

Notice that this definition says nothing about the order of the elements. Hence the elements of two equal sets do not have to be in the same order, it is sufficient that they only have (exactly) the same elements. For the sake of clarity and ease of describing sets, however, we will continue to order them.

By this definition, sets Q and R in Table 2 (above) are equal because every element of Q is contained in R, and every element of R is contained in Q; hence, we may write $Q = R$. They are clearly one and the same set, they just have different symbolic representatives.

Besides equality between sets, we can also compare whether a set is a *proper subset* of another set. In the case of our set of silverware pieces, the set that contains all the spoons and only the spoons is a proper subset of the set of silverware pieces. The set of just the forks is also a proper subset of the set of silverware pieces. We define a proper subset as follows:

∘ DEF. 1.3.2: PROPER SUBSET

Set B is a proper subset of a set A, if and only if every element of set B is an element of set A and there is at least one element of A that is not an element of B. We indicate the relation of being a proper subset by writing $B \subset A$ (read B is a proper subset of A). We also define $A \supset B$, as A is a proper <u>superset</u> of B, if and only if $B \subset A$.

Because every set is either a proper subset of itself *or* equal to itself (It is *equal* to itself!), we may write $A \subseteq A$, in much the same fashion that we would write $7 \leq 7$; and if a set, say B, is a proper subset of some other set, A, (i.e., $B \subset A$), then it is certainly the case that either $B \subset A$ *or* $B = A$ (in this case we have $B \subset A$); we may, therefore, write $B \subseteq A$, in much the same way that we would write $5 \leq 7$. We read the symbol, $\subseteq$, as "*is a proper subset of or equal to*", while the symbol, $\supseteq$, is read "*is a proper superset of or equal to*".

On the other hand, there are some sets in Table 2 that have no elements in common. Set W has no element in common with any of the sets T, Q, R, U, or V, hence W is considered *disjoint* with respect to each of them, and each of them, of course, have no elements in common with W, hence they are disjoint with respect to W. There is no *relational* symbol for indicating that two sets are disjoint; we simply indicate it in English.

Finally, there are some sets, like U and V in Table 2, that are not proper subsets, are not equal, and are not disjoint. In this case, they are just a pair of sets that have some elements in common and some not.

Besides comparing sets, there are two important operations associated with sets that are analogous to the words *and* and *or* in English and logic. They are the *intersection,* $\cap$, of sets, and the *union,* $\cup$, of sets respectively. The result of either operation is another set.* Thus, $A \cap B$ and $A \cup B$ are sets.

Venn diagrams are also often used when talking about sets, subsets, intersections, unions, etc. In Figure 3, there are two sets, A and B, characterized by circles that have some alphabet letters as elements enclosed within. As you can see they have one element in common, the letter *c*.

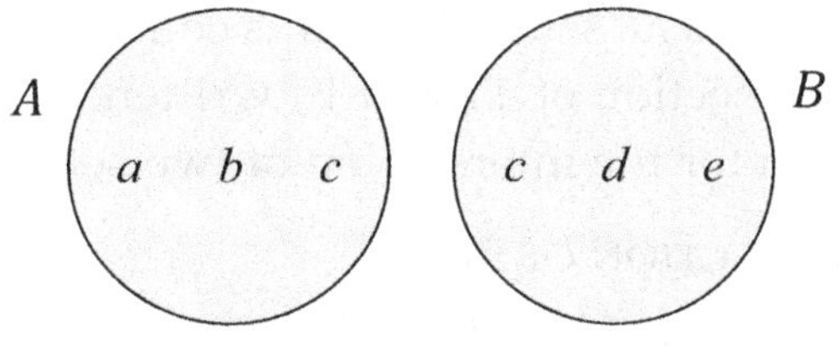

Figure 3: Venn Diagrams – two sets containing alphabet letters.

The intersection of two sets (Figure 4), as the name implies, is a set that contains only those elements common to both (just as two lines intersect at a common point—although sets may have more than one "point" in common). Thus, the intersection of our two sets, written $A \cap B$, which we will call set C, consists of, in this case, the single element *c*. (Note that the element *c*, in lowercase, is not the set C.)

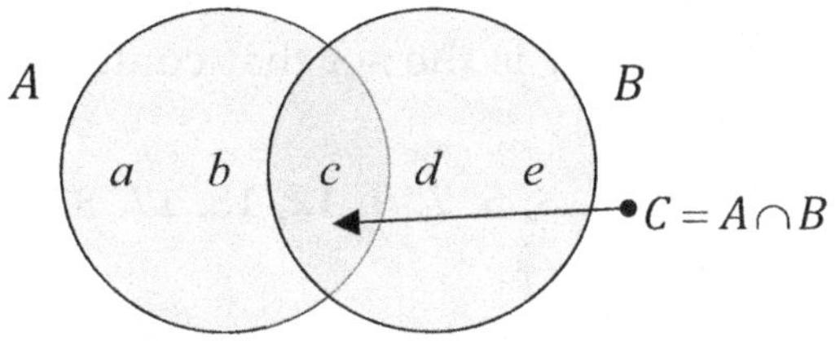

Figure 4: Venn Diagrams – the intersection is represented by the darker (middle) shaded area.

The union of two sets (Figure 5) is a set that contains all the elements from both sets. Thus, the union of our two sets, written $A \cup B$, which we will call set D, consists of all the elements *a, b, c, d,* and *e*. (The element *c* is not repeated.)

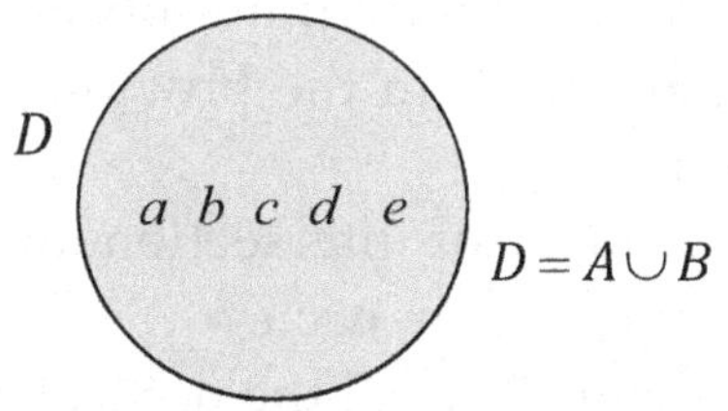

Figure 5: Venn Diagrams – the union of the two sets.

These ideas of the intersection and the union of sets correspond nicely with our notions of the words *and* and *or,* especially when we are talking about

* By analogy, the operation of addition yields another number.

numbers. Consider the sets T and V again from Table 2 (page 27), and suppose that, in the first case, we would like to choose a number from T *and* V, and in the second case we would like to choose a number from T *or* V. What should, or could, our choices be in each case?

In the first case, we should not choose any of the numbers from among 1, 2, 3, 7, 9, 12, 15, 34 and 49 because these numbers only be long to one or the other of the sets and not to both. To say that the number 2 is an element of T *and* V seems nonsensical. To make sense, we need to choose an element that is common to both T and V. Hence, it seems reasonable to choose any of the numbers 5, 17 or 81. The set that has this collection of common elements is derived from the intersection of T with V, written $T \cap V = \{5, 17, 81\}$.

The formal definition for the intersection of two sets is as follows:

∘ DEF. 1.3.3: THE INTERSECTION OF SETS

$x \in A \cap B$, if and only if $x \in A$ and $x \in B$. (This would be read as, x is an element of the intersection of A and B, if and only if x is an element of A and x is an element of B.)

It should be apparent that regardless of which set comes first, the result of an intersection will be the same, that is to say, $A \cap B = B \cap A$.

In the second case, where we are to choose an element from either T *or* V, it seems reasonable to choose any of the numbers from either set. To say that 2 (or 1, 3, 5, etc.) is an element of T *or* V seems quite reasonable. Hence the union of T with V, written $T \cup V$, is the set that contains the elements from both sets (again, less any repeats):

$$T \cup V = \{1, 2, 3, 5, 7, 9, 12, 15, 17, 34, 49, 81\}.$$

∘ DEF. 1.3.4: THE UNION OF SETS

$x \in A \cup B$ if and only if $x \in A$ or $x \in B$. (This would be read as, x is an element of the union of A and B, if and only if x is an element of A or x is an element of B.)

It should be apparent here as well that regardless of which set comes first, the result of a union will be the same, that is to say, $A \cup B = B \cup A$.

The use of the word *or* in this definition behaves in the same fashion as it did when we were discussing it in the previous section on logic: "or" means one or the other, or both.

To see how these two ideas of intersection and union work in algebra, we shall consider a case example from each.

The Case for Intersection: Find all the natural numbers, x, such that $x > 15$ *and* $x \leq 21$. The use of the word *and* indicates that both of these conditions on x *must* be satisfied. The first part of the *and*-statement can be understood as, "x is a natural number greater than 15". This implies the set of natural numbers from 16 to infinity: $\{16, 17, 18, \cdots\}$. The second part states that "x is less than or equal to 21". This implies the set of natural numbers between 1 and

21 inclusively: $\{1, 2, 3, \cdots 20, 21\}$. The conjunction *and* is indicative of the intersection of these two sets, which is the set of elements common to both:

$$\{16, 17, 18, \cdots\} \cap \{1, 2, 3, \cdots 20, 21\} = \{16, 17, 18, \cdots 20, 21\}.$$

Hence, any value for x taken from the set $\{16, 17, 18, \cdots 20, 21\}$ will satisfy all three of the conditions placed on x: (1) x is a natural number, (2) $x > 15$, *and* (3) $x \leq 21$. Specifically, 19 is one of the natural numbers in the set—it is greater than 15 *and* less than 21, while 3 is not in the set—it is only less than 21.

<u>The Case for Union</u>: Find all the natural numbers, x, such that $x > 15$ *or* $x \leq 21$. The use of the word *or* indicates that either (or both) of these conditions on x *may* apply (but not neither). The first part of the *or*-statement can be understood as, "x is a natural number greater than 15". This implies the set of natural numbers from 16 to infinity: $\{16, 17, 18, \cdots\}$. The second part states that x is less than or equal to 21. Again, this implies the set of natural numbers between 1 and 21 inclusively: $\{1, 2, 3, \cdots 20, 21\}$. However, the union of these two sets will consist of all those elements in the set $\{16, 17, 18, \cdots\}$ *or* in the set $\{1, 2, 3, \cdots 20, 21\}$, which is just the set of all the elements from both sets, less any repeats:

$$\{16, 17, 18, \cdots\} \cup \{1, 2, 3, \cdots 20, 21\} = \{1, 2, 3, \cdots 16, 17, 18, \cdots 20, 21, \cdots\}.$$

This is the set that satisfies all three conditions on x and is, as it turns out in this case, the complete set of natural numbers. To see that this is indeed so, we can pick a number from the set and test it, say the number 10,332. Well, the number 10,332 is certainly a natural number greater than 15 *or* less than 21 —it is *greater than* 15. Likewise, 3 is a natural number greater than 15 *or* less than 21 —it is *less than* 21.

With all that has been said, there is still at least one more question that has not been answered: what are we to make of the intersection of two sets that have no elements in common, the disjoint sets that were mentioned earlier? On carefully examining the sets from Table 2 (page 27) again, it becomes apparent that the set V and W (among others) have no elements in common:

$$V \cap W = \{1, 5, 15, 17, 49, 81\} \cap \{8, 13, 19, 67, 95\}. \qquad ▪1.3.1$$

Hence the result of their intersection should be the set that contains no elements: { }; and in fact, it is. Nothing but endless sleep without dream or the number 0 comes closer to expressing the notion of *nothing* than the "contents" of the *empty set*. The empty set is also known as the *null set*, and it is often written as either $\emptyset$ or { }; hence, $V \cap W = \emptyset = \{\}$. Although the set itself is something, it contains nothing. Hence, the set that contains zero (i.e., $\{0\}$), is not empty. (The mind of a mathematician is a mysterious thing.)

The idea of the null set is indispensable to algebra. For example, we might wish to know what set of natural numbers, x, will satisfy both $x \leq 77$ and

$x > 102$. It should be clear that there is no natural number, x, that is both less than or equal to 77 *and* greater than 102. In set notation we would write

$$\{1, 2, 3, \cdots 76, 77\} \cap \{103, 104, 105, \cdots\} = \{\ \}.$$

Some of you may have noticed that the second set in the above sentence begins with 103 and not 102; the reason is that the set of all natural numbers *greater* than 102 must begin with the natural number 103. I could have just as well written $x \geq 103$ to say the same thing, which brings us to one other important method for expressing the idea of a set of numbers.

If you recall, the elements of a set do not need to be ordered, I have been ordering them for the sake of clarity and ease of reading, which is always a sensible thing to do. There is, however, a notation that when used always indicates an *ordered* set of numbers. It is called *set-interval notation*.

The best way to present it is with examples, followed by a brief explanation, but be aware, context is important when working with set interval notation. For now, we will only be talking about the natural numbers. However, it works just as well for any ordered and predictable sequence of numbers.*

- *Ex. 1*: $x \geq 7$ and $x < 19$ can be indicated in interval notation as $[7, 19)$.
  - Explanation: the left "square" bracket, which in set theory is called a *closed bracket* or *closed parenthesis*, indicates that the number 7 is included in the set and that it is the least number in the set (the lesser numbers always go to the left, while the greater numbers go to the right). The parenthesis on the right, which is called an *open bracket* or *open parenthesis*, indicates that the number 19 is *not* in the set. Since we are talking about an ordered collection of *natural* numbers, the next natural number less than 19 is 18, and it (not 19) *is* in the set. We say that the interval, $[7, 19)$, is closed on the left and open on the right.
- *Ex. 2*: $x \geq 7$ and $x \leq 18$ can be indicated in interval notation as $[7, 18]$.
  - Explanation: the square brackets on either side, both of which are *closed brackets* or *closed parentheses*, indicate that both numbers are included in the ordered set of natural numbers from 7 to 18. In this case we have $[7, 19) = [7, 18]$. We say that the interval, $[7, 18]$, is a *closed* interval.
- *Ex. 3*: $x > 7$ and $x < 18$ can be indicated in interval notation as $(7, 18)$.
  - Explanation: the parentheses on either side, both of which are *open brackets* or *open parentheses*, indicate that neither number is included in the ordered set of numbers. In this case, we have $(7, 18) = [6, 17]$.† We say that the interval, $(7, 18)$, is an *open* interval.

---

* By that I mean one's *audience* can predict a number from the given sequence of numbers.

† Context is important, and one needs to be careful. When talking about the set of rational numbers, for example, none of the following intervals are equal: $(\sqrt{2}, 3)$, $[\sqrt{2}, 3)$, $[\sqrt{2}, 3]$, $(\sqrt{2}, 3]$.

These two methods for displaying the contents of a set—standard notation and set-interval notation—may be used in combination with one another. Thus, it would be correct to write {7, 8, 9, 10, 11, 12} = [7, 12] even though the two notations have slightly different definitions: the former being unordered, and the latter being strictly ordered. We can do this because whether a set is ordered or unordered, it is still the same set. Consequently, it would be perfectly correct to write {8, 11, 7, 12, 10, 9} = [7,12] as well, and to use the various operations of set theory with both types of notation. For example, it would be correct to write $[7, 12] \cup \{0\} = \{0, 7, 8, 9, 10, 11, 12\}$. However, it would (may) be incorrect to write, in this instance, that $[7, 12] \cup \{0\} = [0, 12]$ because, although the set {0, 7, 8, 9, 10, 11, 12} has been displayed in an orderly fashion, it is not predictive (the gap between 0 and 7 is greater than the gap between the other numbers). Context and making sure your audience knows precisely what it is you are discussing is what is most important.

Our discussion about sets should not leave you believing that we are restricted to working with only two sets at a time. We may work with any number of sets and apply any of the concepts we have discussed as our needs require. For example, Figure 6 (below) shows three sets, A, B and C, each characterized, again, by circles, and where each has (again) alphabet letters within. The sets are made up of the elements a, b, c, d, e, f, and g as follows: $A = \{a, d, e, g\}$, $B = \{b, d, f, g\}$ and $C = \{c, e, f, g\}$.

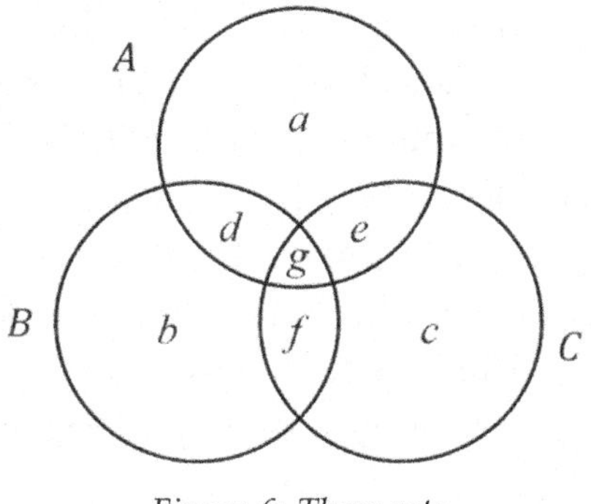

Figure 6: Three sets

When working with three or more sets it is important to use parentheses to keep the correct order in which the operations of union and intersection are to be performed, just like in arithmetic. Hence, $(A \cap B) \cup C = \{c, d, e, f, g\}$ because the parentheses demand that the intersection of A and B be performed first, thus obtaining $A \cap B = \{d, g\}$, after which we perform the union of that set with the set C to obtain $\{c, d, e, f, g\}$. If we were to shift the parentheses to the right and perform the union first, we would have obtained $A \cap (B \cup C) = \{d, g, e\}$, and clearly $(A \cap B) \cup C \neq A \cap (B \cup C)$.

Finally, if all of this were not enough, we can also introduce the idea of the *complement* of a set. The complement of a set A is the set denoted by A' (read

A prime[*]), which is the set that contains all of the elements that are *not* in the set A. Thus (turning our gaze to Figure 6 again), because $A = \{a, d, g, e\}$, it follows that the complement of A is $A' = \{b, f, c\}$, and, as a side note, we have $A \cap A' = \{\ \}$ (the empty set), and $A \cup A' = \{a, b, c, d, e, f, g\}$ (the set of *all* of the elements under discussion).

Taking the complement of set is considered a *unary* operation because it operates on one set, while union and intersection are considered *binary* operations because two sets are required; that is to say, A' makes sense, but $A \cup$ and $A \cap$ are nonsensical.

We now have the tools necessary to construct a set containing some, none or all of the elements from the three sets shown in Figure 6. For example, to obtain the set $\{a, c\}$ we would do the following, $(A \cup B)' \cup (B \cup C)'$. In other words, we first find the complement of the union of A with B, then find the complement of the union of B with C, after which we take those two sets and perform a union with them and, voilà, we have the desired set, $\{a, c\}$.

There are 128 possible sets (including the null set, which happens to be a subset of every set) that can be constructed from the three sets shown in Figure 6 using the ideas of union, intersection and complement of a set. I will provide one more example, and then leave it for you as a personal exercise to determine how to arrive at some of the other possible sets†

For this last example, I have chosen to construct the set $\{d, f\}$. To arrive at it we can do the following: $((A \cap B) \cup (B \cap C)) \cap (A \cap B \cap C)'$. It is up to you to convince yourself that this does indeed equal $\{d, f\}$.‡ (☆2)

This leaves us with some unanswered questions. Was the method that I used to arrive at the desired set, $\{d, f\}$, which required seven operations, the method that required the least number of operations? Is there a general rule for finding the most efficient method for constructing any desired set from three or more sets? And, did somebody else already answer these questions? These are the things that gnaw at a mathematician's peace of mind.

~

* Often the superscript "C" is used instead of the prime symbol.

† For those of you who are interested and know a little bit about mathematical combinations (or even if you don't), I leave it to you to demonstrate that the number of sets derivable from Figure 6 using the operations of union, intersection, and complement is equal to the number of subsets that can be composed from a single set with seven elements.

‡ The operations in parentheses are performed first.

CHAPTER 1 EXERCISES

1. Intuition can often fail us. For example, before Albert Einstein published his works on relativity, every person believed in their soul that time "ticked" away at the same pace for everyone and for everything, everywhere across the universe—God's clock, so to speak. We know now that that's not true. What does your intuition tell you about there being an *exact* formula for finding prime numbers? What does it tell you about mathematics and its relation to the universe? What does your intuition tell you about the likelihood of there being aliens from another planet? What does it tell you about the possibility that humans will eventually create artificial intelligence? For each of these questions, ask yourself why you believe the way you do.
2. Axioms have the property of being self-evident (but they don't have to be) and the property of being mandates (they *always* have this property), or laws if you will, that cannot be broken. Suppose the statement, "If Nancy runs fast, then pigeons can't fly" is an axiom. What, then, can be known for certain from the statement, "Nancy runs fast"?
3. *If you read PM and you eat brains, then you are a psycho-nerd or a zombie.*
    a. If the above conditional represents the original, write the three other conditionals and label them accordingly.
    b. If the above conditional represents the inverse, write the three other conditionals and label them accordingly.
    c. For problem "a" above, indicate which of the conditional statements have equivalent truth values.
    d. Write the given statement as an if-and-only-if statement.
4. What is the negation of each of the following phrases.
    a. It rains and snows.
    b. The sun shines or the moon glows.
    c. The boy likes apples or carrots, and drinks lemonade.
    d. The girl likes spaghetti and meatballs, or doesn't like soda.
5. What is the contrapositive of each of the following conditionals?
    a. If p and q, then $\sim s$ or t.
    b. If $\sim(p$ or $q)$ and s, then t and w.
    c. If $(p$ or $\sim q)$ and $(r$ and $\sim s)$, then $\sim(t$ or $w)$.
    d. If p or $\sim(q$ and $(r$ or $s))$, then w and $(v$ or $x)$.
6. Mathematicians do love their symbols; so much so that they use the symbol $\wedge$ to represent the word *and*, the symbol $\vee$ to represent the word *or*, and the symbol $\Rightarrow$ to represent the word *implies* (i.e., *if. . ., then. . .*). We have already mentioned the symbol $\Leftrightarrow$, which is used to mean that two statements are logically equivalent, which also replaces the words *if, and only if.* Write the following conditionals in symbols (be consistent with your choice

of logical variables):

a. It rains in Spain if, and only if, the plains are dry and clouds have gathered.
b. If the plains are dry and clouds have gathered, then it rains in Spain.
c. If the plains are not dry or clouds have not gathered, then it doesn't rain in Spain.
d. If it doesn't rain in Spain, then clouds have not gathered.

7. You are a detective and you know with absolute certainty that the victim was killed with a rope, or a knife or a Mongolian yak. Your intense and comprehensive investigation reveals that if either the rope or the knife was the cause of death, then your known facts about the case would be contradictory. What must you conclude?
8. After examining the set of Common Silverware Serving Pieces on page 24, describe the set (using the Roman numeral) that contains just knives (you may include the spreaders in this list as well). What is the intersection of this set with the set of all those pieces that contain the word *butter*? What is the union of these two sets?
9. Using the three sets of Figure 6 on page 33, determine the results of the following set operations.

a. $((A \cup B)' \cap C')'$ b. $(A \cap B \cap C)'$ c. $(A \cap B) \cup C$
d. $(B')'$ e. $A \cap (B \cup C)$ f. $A \cap \varnothing$

10. Determine the truth or falsehood of the following statements using Figure 6 on page 33; if a statement is ill-formed, say so and explain why.

a. $B \cap C \subseteq B \cup C$ b. $d \in B \cup C$ c. $A \cap B \cap C = g$
d. $A \cap \{b, f, c\} = \varnothing$ e. $\{d, g, e, f\} = \{a, b, c\}'$
f. $\{d, g, e, f\}' \neq \{a, b, c\}$ g. $B \in A \cup B \cup C$

11. How many unique subsets can be constructed from a set that contains three elements? Four elements? Five and six elements? Find a formula for any given number of elements (*Remember: the null set is a subset of every set*).
12. Consider the following two sentences: (1) If I were a bird, then I would have feathers; (2) If I were a bird, then I would have *blue* feathers. Why is one sentence considered logically "proper" and the other not? How does this relate to logical variables such as x and y?

CHAPTER 2
THE AXIOMS OF COMPARISON

Axioms are the embryonic statements from which a mathematical subject is created. They are considered indomitable truths, or mandates, if you will, and do not require proof. When a collection of axioms come together it is considered an axiomatic system. Axioms "act" against and with one another to produce theorems, which are just the logical consequences of these interactions.

The most famous axiomatic system is the one developed by Euclid (circa 300 B.C.), which he used to create the monumental mathematical structure known as *The Elements of Geometry*. In that work, Euclid employs five axioms from which all other statements would logically follow.* One logically deducible statement was Pythagoras' theorem: the area of the square on the hypotenuse of a right triangle (square C) is equal to the sum of the area of the two smaller squares on the two legs of a right triangle (i.e., the sum of the areas of squares A and B).

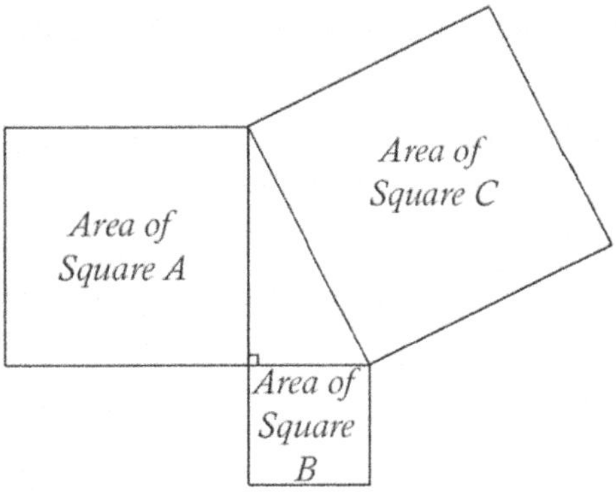

Axiomatic systems have three properties: *consistency*, *independence* and *completeness*. An Axiomatic system is consistent if not both a theorem and its negation can be proved from the axioms; in other words, it cannot contradict itself. For example, if we can prove that $1 \neq 0$ *and* $1=0$, then the system of axioms is *inconsistent*. An Axiomatic system is independent if no axiom can be proved from any combination of the other axioms. Thus, if we can prove, say, Axiom#5 using Axiom#2 and Axiom#4, then the system of axioms is not independent. In such a case, Axiom#5 would be better identified as a theorem. Finally, an axiomatic system is complete if we have all of the axioms necessary to prove or disprove every statement that one can make (which cannot be known beforehand) regarding the subject to which the axioms pertain. Thus, if algebra did not have the distributive axiom, we could not prove that $3\times 0=0$.

Although it is often possible to directly show whether a system of axioms is

---

* Mathematicians would later conclude that closer to 20 axioms are required to logically create Euclid's plane geometry, although to his credit, he had considered some of these as "postulates".

independent by examining the axioms themselves, it is quite another matter to show that the system is both consistent and complete.

Algebra (for our purposes) is composed of nineteen axioms, ten of which are about the comparison of numbers (the subject of this chapter) and nine of which (covered in Chapters 3 and 4) are about the properties that all numbers have in common.

Although it is not the case that a system of axioms must represent or model "reality" (i.e., one's imagination is free to wonder), in the case of algebra they do: they model the ancient history of numbers.

2.1 EQUALITY

We all have a general notion about what equality means. We often speak of social equality, judicial equality or economic equality to name a few. However, unlike these vague notions, equality in mathematics has a very precise meaning. In mathematics, whenever we declare that two objects are equal, regardless of which branch of mathematics they come from—logic, sets, algebra or what have you—we mean that the two objects must abide by the axioms of equality as outlined below, or at the very least the first three, otherwise the two objects are decidedly not equal, regardless of any definition claiming otherwise. However, before I continue along this vein, allow me to first present the axioms of equality and briefly explain not only why, from a logical and symbolic point of view, they are necessary, but also a few of their limitations.

❖ **THE SIX AXIOMS OF EQUALITY**

For any numbers a, b, *and* c

(E1) $a=a$	Reflexive Property
(E2) *If* $a=b$, *then* $b=a$	Symmetric Property
(E3) *If* $a=b$ *and* $b=c$, *then* $a=c$	Transitive Property
(E4) *If* $a=b$, *then* $a+c=b+c$ *and* $c+a=c+b$	Addition Property
(E5) *If* $a=b$, *then* $a\cdot c=b\cdot c$ *and* $c\cdot a=c\cdot b$	Multiplication Property
(E6) *If* $a=b$, *then* a *may replace* b *(or* b *may replace* a*) in any expression, formula or statement that contains* b *(or ... that contains* a*)*	Substitution Property

(The numbers, a, b, and c, although *visually* distinct, do not preclude the possibility that they could be the same number: 7, 6 and 6, or 5, 5, and 5 are *any* numbers even though some or all of them are the same number.)

We should not read anything else into these axioms, or so say the math formalists* anyway. The axioms are what they are, and these are just the first six of the nineteen fundamental laws of algebra that kick start the entire process of deduction. Furthermore, the word *number*, as used in the axioms, is just

* Formalists believe that mathematics is the strict manipulation of symbols based on rules.

that, a word and nothing more. It too is technically without an interpretation.

The reflexive axiom is needed because there is no other way of deducing the statement $a=a$; consequently, it must be stated explicitly. Try as you like to deduce this statement from the other five axioms and all you will be able to arrive at is a headache. However, without it we would not, logically, be able to arrive at what we know to be an absolute certainty from an interpretive point of view, and that is that every number we have ever had the pleasure of encountering is equal to itself, $a=a$, $5=5$ and $3.14=3.14$ for example. We would only be able to arrive at the lesser statement "*If* $a=a$ ..." or "*if* $5=5$ ..." from the remaining axioms of equality, which is totally unsatisfactory.

There is one other important reason for including this axiom: it allows us to make a distinction between the equality and the inequality symbols, which we will consider in more detail when we get to the axioms of inequalities.

Likewise, the remaining axioms of equality cannot (arguably) be deduced from any creative combination of the other axioms*, but they too are essential if our axioms of equality are going to behave in a way that we know equality between numbers should behave. They are considered self-evident and were chosen in large part to coincide with our fundamental beliefs about equality (but were primarily chosen to provide the necessary and sufficient conditions to logically construct not only algebra but arithmetic as well).

However, from a strictly logical and symbolic point of view, they are no more self-evident than the following it's-all-Greek-to-me statement:

If $\sqcup$, $\Im$, *and* $\wp$ *are doodads and if* $\sqcup \Updownarrow \Im$, *then* $\sqcup \because \wp \Updownarrow \Im \because \wp$. ▪2.1.1

It is only because of our familiarity with the symbols of arithmetic that the symbols used in the axioms of equality allow us to perceive the self-evidentiary nature of the axioms, while, in contrast, those in 2.1.1 convey little to no information at all, yet the visual form of 2.1.1 is exactly like that of Axiom E4.

That brings us to our next point: the precise form or arrangement of the symbols in an axiom must be strictly obeyed, whenever there is no logical reason that would prohibit us from doing otherwise. For example, Axiom E4 does not allow us to write "*If* $a=b$, *then* $a+c=c+b$", where the position of c in relation to the $+$ symbol has been swapped—there is no logical reason that will allow it (yet). However, Axiom E4 does not prohibit us from writing $a+c=a+c$ because there is no rule restricting us from using the same number in any of the axioms of equality. Unless there is a logical reason (axiom or theorem) prohibiting us from using the same number, we are free to do so. Each axiom limits what we may or may not write based strictly upon its form and what each logically implies or doesn't logically prohibit.

Furthermore, and this is critical, neither Axiom E4 nor E5 tell us how to

---

* However, one may try to argue that some of these axioms are deducible from Axiom E6.

construct or obtain any of $a+c$, $b+c$, $a\cdot c$ or $b\cdot c$, or even what these strings of symbols mean, or even if they are numbers. (Comparatively, the conditional statement in 2.1.1 does not tell us what $\amalg\because\wp$ or $\Im\because\wp$ mean either.) Although the axioms of equality do not explain for us what + or · mean or how to obtain such things as $a+c$, $b+c$, $a\cdot c$ or $b\cdot c$ other axioms do.

Also, Axioms E4 and E5 are not a strict requirement for equality in every branch of mathematics (although it is for algebra) because not every branch of mathematics uses the ideas of addition or multiplication. Sets, for example, do not use either. However, the first three properties are the hallmark of equality and are, as previously mentioned in the introduction to this section, a necessary requirement whenever any two mathematical objects are said to be equal, regardless of the branch of mathematics from which they come.

For example, suppose we wish to discuss lines in a plane and we have decided that equality between lines shall be defined as *two lines are equal if and only if they are parallel.** This will be a legitimate definition of equality between any pair of lines provided that this definition can be shown to be reflexive, symmetric and transitive.

In the first case, it should be obvious that every line is parallel to itself. Hence the definition conforms to the reflexive property. In the second case, if a line, l_1, is parallel to another line, l_2 (written as $l_1 \,||\, l_2$), then it is certainly true that line l_2 is parallel to line l_1 ($l_2 \,||\, l_1$). Hence, the definition conforms to the symmetric property. Finally, if two lines are each parallel to a third line, then it is also true that the two lines are parallel to each other (i.e., if $l_1 \,||\, l_2$ and $l_2 \,||\, l_3$, then $l_1 \,||\, l_3$). Hence, the definition conforms to the transitive property.

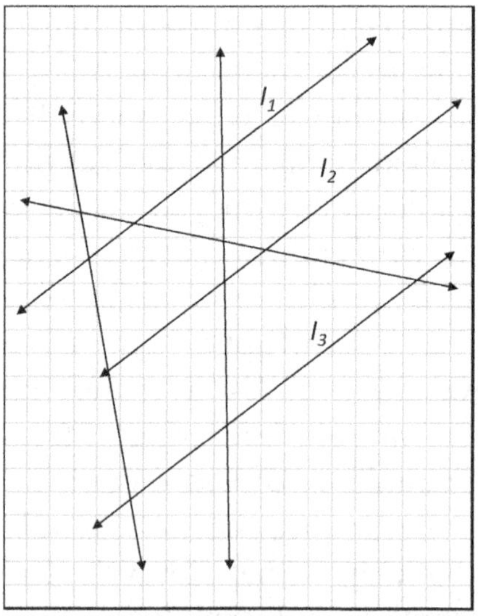

Figure 7

* Definitions should always be written with if-and-only-if statements, but many (lazy or careless) mathematicians neglect to write the "and-only-if" part, so beware.

Thus, it makes perfectly good sense to write $l_1 = l_2$ when speaking about lines in this context.

Recalling our definition (Def. 1.3.1) of equality between sets from the previous chapter, we can now test if that was a good or poor definition by determining whether it is or is not reflexive, symmetric and transitive.

The reflexive property holds because if A, B, and C are sets and if $x \in A$, then (obviously) $x \in A$; consequently, $A = A$. (Recall that prior to the middle of the nineteenth century, a deduction having only one premise was not an appropriate deduction.)

The symmetric property holds because $A = B$ means, by definition, "if $x \in A$, then $x \in B$, *and* if $x \in B$, then $x \in A$." However, the word *and* in this definition is symmetric, so we can also write the equivalent statement "if $x \in B$, then $x \in A$, and if $x \in A$, then $x \in B$," but this is the definition for $B = A$; consequently, $A = B$ implies $B = A$.

Finally, the transitive property holds because if $A = B$ and $B = C$, then $x \in A$ implies $x \in B$; however, $B = C$, so $x \in B$ implies $x \in C$, therefore $x \in A$ implies $x \in C$. Because we have already shown that the symmetric property holds, we can, in like fashion, show that $x \in C$ implies $x \in A$. Putting it all together, we have $A = C$ by Def. 1.3.1. Consequently, $A = B$ and $B = C$ implies $A = C$.

If all of this seems rather trivial, I apologize, but the symbol "$=$" cannot be used willy-nilly in any mathematical context. If you are comparing mathematical objects of your own creation (and my wish is for you to create your own mathematics) and you cannot show that these three basic properties of equality hold for that comparison, then you cannot (should not) use the equality symbol for that comparison, but that does not mean that you cannot devise a new symbol for the comparison you have in mind.

A brief comment should be made about Axiom E6, if for no other reason, then because it is a good bit wordier than the other axioms. It states that if we are given an expression like $3a + 7$ (expressions are sentence fragments), and a statement like $a = (2/9)b + 2$ (statements, formulas or equations are complete sentences) then we may write $3a + 7 = 3[(2/9)b + 2] + 7$, where the single occurrence of a in the expression was replaced with its equal from the statement. (A word of caution: always enclose the number that is replacing the original number with parentheses before you make the exchange. This will help reduce copying errors. In this case, I have enclosed $(2/9)b + 2$ within a pair of square brackets, i.e., $a = [(2/9)b + 2]$, *before* I made the replacement.)

The mathematical definition of equality to the uninitiated may appear trivial and obvious but its study and utility in such diverse fields of mathematics as number theory and group theory has wrought powerful abstract mathematical ideas.

~

2.2 INEQUALITY

Numbers, as we know them, are ordered from lesser to greater. And it is obviously clear which numbers are lesser and which are greater when our discussion is restricted to just the natural numbers, $N = \{1, 2, 3, \cdots\}$. Thus 2 is less than 3; and we express the idea of this lesser-than relation mathematically as $2<3$. The axioms of inequality, or the defining characteristics of the symbol "<", are as follows:

❖ **THE FOUR AXIOMS OF INEQUALITY (THE ORDER AXIOMS)**

For any numbers a, b, and c

(**O1**) *Exactly one of the following conditions must hold for any two numbers:*

$a=b$, $a<b$ *or* $b<a$ — Law of Trichotomy

(**O2**) *If* $a<b$ *and* $b<c$, *then* $a<c$ — Transitive Property

(**O3**) *If* $a<b$, *then* $a+c<b+c$ — Addition Property

(**O4**) *If* $a<b$ *and* $0<c$,* *then* $a \cdot c<b \cdot c$ — Multiplication Property

And we define " > ", which is read as *is greater than,* as $a>b$ if and only if $b<a$. (As with the axioms of equality, the numbers a, b, and c do not, *necessarily*, preclude the possibility that they are all the same number.)

Please note that whenever mathematicians introduce a new definition, they are often either explaining how they intend to use a new, hitherto (supposedly) unfamiliar symbol for the purposes of brevity and clarity, or how they intend to use a familiar symbol differently. At the dawn of calculus, for example, the symbol $\int$ (an elongated S introduced by the German mathematician Leibniz) was introduced to indicate the novel idea of a continuous sum, which can be likened to summing *all* the numbers along a line segment, as opposed to a discrete sum, which involves numbers like the integers.†

In our case, the greater-than and less-than symbols, where the greater-than symbol is defined in terms of the less-than symbol, were introduced at about the same time in the 16th century. However, both are not needed to do mathematics; it is possible to perform all of mathematics by just using the less-than symbol, but the mathematics would be cumbersome and difficult to follow.

In fact, all of mathematics can be performed without using any symbols at all, just as it was done, for the most part, some centuries ago. The only conditions placed on mathematics are the ability to communicate one's thoughts and ideas, and a logically compelling argument; beyond that symbols are only needed for brevity and clarity (although, to the uninitiated it would seem to be just the opposite).

You will notice that there is neither a reflexive axiom nor a symmetric axiom

---

* However, these axioms have not told us what 0 is, but Axiom A4 does—patience.

† Analogously, it is the difference between measuring the *mathematically exact* length of a string and counting the number of marbles in a bag.

in the list of axioms for inequality like there was for the axioms of equality. That's because from an interpretive—intuitive—point of view a number cannot be greater or less than itself, and a number cannot be greater than some number and simultaneously less than that number.

From a logical point of view, however, if the less-than symbol had both of these properties including the transitive property, which it already has, then it would in fact be a substitute symbol for the equality symbol, because all three of these properties taken together are strictly reserved for our idea of equality. In other words, if $<$ had the additional properties of $a<a$ (reflexive), and $a<b$ *implies* $b<a$ (symmetric), then $<$ would behave precisely like $=$ behaves, and two different symbols behaving exactly alike would not only be unnecessary but also confusing. Furthermore, if there were either a reflexive or a symmetric axiom for the inequality symbol, then some of our axioms would contradict each other.

Specifically, using the axioms of equality and inequality in concert with one another, we can prove that $a<a$ * is false quite simply as follows: the reflexive axiom of equality (E1) states that $a=a$, and the law of trichotomy (O1) states that one and only one of three conditions may apply. In this case, because $a=a$ applies, it follows that $a<a$ does not apply and therefore is false. Consequently, the reflexive property does not hold for inequalities.† Hence, having a reflexive axiom for $<$ would have led to a contradiction in the axioms. Thus, we cannot have the same number on either side of $<$, *logically* speaking.

In a similar fashion, we can prove that the symmetric property fails for $<$ as well. For suppose that $a<b$, then by the law of trichotomy $b<a$ is false (as is $a=b$). Hence the relation symbol $<$ is not symmetric‡.

We may also define the symbol $\leq$, as in $a\leq b$, to mean either $a<b$ or $a=b$. Likewise, we define $\geq$, as in $a\geq b$, to mean $a>b$ or $a=b$. Thus, $5\leq 7$ and $7\geq 5$.

There is another very useful and logical result we can derive from the law of trichotomy, the truth of which we naturally accept because of our intuitive understanding of numbers, and that is if we can eliminate one of the three relations that could exist between a pair of numbers, then, according to the law of trichotomy, the relation between the numbers must be one of the remaining two. Thus if $a\nless b$ §, then it must be the case that $a=b$ or $a>b$, which we could write more concisely as $a\geq b$; in arithmetical terms, if $7\nless 5$, then it must be the case that either $7=5$ or $7>5$. likewise, if $a\neq b$ then it must be the case that

---

* Reflexive, from a structural perspective, means that the same symbol may appear on either side of some central symbol, thus the symbol "□" as used in p □ p is reflexive.

† This gives another reason for the importance of having a reflexive axiom of equality because without it we could not prove that 5 < 5 is false.

‡ Symmetry, from a semantic perspective, means that the symbols on either side of a central symbol may be swapped, thus the symbol ✣ as used in x ✣ p and p ✣ x is symmetric.

§ A line though a symbol means the negation of the symbol, thus ≠ means "*not* equal to", etc.

$a<b$ or $b<a$. (There is no alternative symbol besides $\neq$ in mathematics to describe the condition of $a<b$ or $b<a$, but because of the Law of Trichotomy that is quite good enough.)

That brings us to our first, formal, theorem, the extreme importance of which will be seen when we begin our discussion on radicals in Chapter 5. The proof is by contradiction.

◆ THEOREM 2.2.1

Given numbers a and b, $0<a$, $0<b$, then $a \cdot a = b \cdot b$, if and only if $a=b$

Proof: Suppose instead that $a \neq b$, then, by the law of trichotomy, it must be the case that either $a<b$ or $b<a$. If $a<b$, then by Axiom O4 (because $0<a$) we have $a \cdot a < a \cdot b$. By that same axiom (because $0<b$) we also have $a \cdot b < b \cdot b$. It then follows from Axiom O2, the transitive property of inequalities, that $a \cdot a < b \cdot b$, but this contradicts our premise, which states $a \cdot a = b \cdot b$. On the other hand, if $b<a$, then, following the same line of reasoning, the result would be $b \cdot b < a \cdot a$, which also contradicts $a \cdot a = b \cdot b$. Ergo, it cannot be the case that $a \neq b$; thus, we must conclude that $a=b$ (there is nothing "outside" of the axioms, thus we have no choice but to make this conclusion). On the other hand, if $a=b$, then $aa=ab$ and by substitution $aa=bb$. Thus, the converse is also true.

This theorem would work equally well if $a<0$ and $b<0$ because in that case we would only have to rewrite these inequalities as $0<-a$ and $0<-b$, whereupon, having thus satisfied the first condition of the premise to Theorem 2.2.1, the second condition, $a \cdot a = b \cdot b$, which is equivalent to $(-a)(-a)=(-b)(-b)$, would then yield the conclusion $-a=-b$. However, $-a=-b$ is equivalent to $a=b$. (The converse is trivially true.) Thus, we have shown that Theorem 2.2.1 is also true when $a<0$ and $b<0$.

However, this theorem would fail if one of the numbers was less than zero and the other was greater than zero because a number less than zero cannot equal a number greater than zero. (Although we have not touched upon the concept of additive inverses, I trust that most of my readers are familiar with the concept and understand how I arrived at these conclusions; if not, then may I suggest you return to this theorem after having completed Chapter 4.)

This theorem exemplifies the distinction between understanding mathematics via intuition and understanding it via logic. Those of you who are more intuitive may view this theorem as obvious and even trivial because in your mind's eye you can "see" that for any given positive number there can only be exactly one number between zero and the given number such that when it is multiplied by itself it equals the given number. On the other hand, those of you who are more logical may view this theorem not only as a logically correct deduction from the axioms but also as a theorem that necessarily *had* to be deduced to ensure there exists one and only one such number that meets the requirement.

We could extend Theorem 2.2.1 to include the equality between the product of any number of identical numbers and the product of other identical numbers of the same count.*

◆ **THEOREM 2.2.1, EXTENSION**

Given numbers a and b, $0<a$, $0<b$, *then* $a\cdot a\ \cdots\ a=b\cdot b\ \cdots\ b$ *(where the number of a factors equals the number of b factors), if and only if* $a=b$

To get a feel for how this theorem can be proved in general (that is for any number of factors), we will verify it (by way of contradiction) for the case when there are three factors, however, you will be required to fill in the blanks.

So suppose $0<a$, $0<b$, $a\cdot a\cdot a=b\cdot b\cdot b$, and that $a\neq b$, then either (1) $a<b$ or (2) $b<a$. Regarding the first inequality (1), $a<b$ implies $a\cdot a<a\cdot b$, which implies $a\cdot a\cdot a<a\cdot a\cdot b$. However, $a\cdot a<a\cdot b$ also implies $a\cdot a\cdot b<a\cdot b\cdot b$. Furthermore, $a<b$ implies $a\cdot b<b\cdot b$ as well, which implies $a\cdot b\cdot b<b\cdot b\cdot b$. It therefore follows from the transitive property of inequalities that $a\cdot a\cdot a<b\cdot b\cdot b$, which, of course, contradicts our original assumption that $a\cdot a\cdot a=b\cdot b\cdot b$. The remainder of the proof, the part where $b<a$ and the converse of this theorem, is yours to consider, and I urge you to verify the extension to Theorem 2.2.1 once again for when there are four, six, and eight factors (☆3—*hint: find the patterns*). We will forgo the general proof of this theorem.

It is important to mention that because Axiom O4 is used to some extent in later chapters, we will need to be thoroughly familiar with it. On that account, we will need to take a moment to examine its form. For your convenience, I have repeated it here:

If $a<b$ *and* $0<c$, *then* $a\cdot c<b\cdot c$.

To use this axiom, we must satisfy both conditions of its premise. Specifically, we must have a pair of numbers, a and b, where one of the numbers is less than the other. That condition is easy enough to satisfy provided the two numbers are not equal. However, for the other condition to be satisfied, we must have a number c such that $0<c$, which may not be immediately available. In other words, if we have a number $c<0$, then the condition is *not* satisfied. However, Axiom O3 allows us to add the number $-c$ to both sides of this inequality and arrive at $0<-c$, which does satisfy this condition of the premise. (However, we will need Axioms A4 and A5 to accomplish this.) We will use this idea of adding a number to both sides of an inequality to arrive at a premise that satisfies all of the conditions of Axiom O4 whenever the need arises.

The axioms of algebra were formulated through centuries of struggle to guarantee that the symbols of algebra will logically behave according to our

* The proof follows almost immediately when the number of identical numbers is some power of 2 (i.e., 2 multiplied by itself some number of times), which I leave for you to consider as an exercise (☆3).

familiar, intuitive and historic interpretation of numbers and that whatever can be shown to be true in algebra can also be shown to be true in arithmetic. It is a good mapping, so to speak, between our understanding of how numbers behave in everyday life, and the symbols and rules we have presented thus far in algebra to logically describe that behavior—a necessary endeavor if mathematicians are ever going to agree on what constitutes algebraic proof.

~

Chapter 2 Exercises

1. Prove that if $a+a=b+b$, then $a=b$.
2. Judging from line 2.1.1 on page 39, the symbol $\Updownarrow$ is analogous to the equality symbol and the symbol $\because$ is analogous to the addition symbol. To which equality axiom would the next sentence be analogous? *For any doodads* $\amalg$, $\Im$, *and* $\wp$, *and* $\amalg \Updownarrow \Im$, *then* $\amalg \odot \wp \Updownarrow \Im \odot \wp$.
3. Suppose we make the claim that two positive integers are equal if and only if when divided by 2 they leave the same remainder. Thus, $8=72$ because when either 8 or 72 is divided by 2 the remainder is 0. Concurrently, $15=121$ because when either of these two numbers is divided by 2 the remainder is 1. Show, or intuitively convince yourself, that this definition of equality between two numbers abides by all the axioms of equality.
4. From problem 2 above, it can be seen that the given definition of equality between two numbers *partitions*, or groups, the set of positive integers into two subsets. One subset consists of all the even, positive integers and the other consists of all the odd, positive integers. Suppose that two positive integers are considered equal if and only if they have the same remainder when divided by 3. (a) Would all the axioms of equality hold in this case? (b) How many partitions of the set of positive integers would there be? (c) Suppose the definition of equality meant that two numbers have the same remainders when divided by 4, would the axioms still hold? (d) How many partitions of the positive integers would there be in this case?
5. Suppose you were shown a glass jar filled with one hundred marbles, evenly distributed between ten different colors, and you were asked to pick out *any* three. Would you be wrong, based upon the request, if you picked out three marbles that were all the same color? Why or why not?
6. The reflexive axiom (E1) of equality is needed because without it we cannot prove that $a<a$ is a false statement, but why do you suppose we need the symmetric property (E2) of equality? After all, the proof that there is no analogous symmetric property for inequalities does not depend on E2, so why must we have it?
7. Answer the following statements with true or false and explain your reasoning or exhibit a counter example which supports your answer. (a) If $a \le b$, then $a<b$; (b) if $a \le b$, then $a=b$; (c) if $a=b$, then $a \le b$; (d) if $5 \le 7$, then $5<7$; (e) if $a \ne b$, then $a<b$ or $a>b$; (f) if $7 \not< 5$, then $7=5$.
8. Show that Theorem 2.2.1 is false if $a<0$ and $0<b$. (*Hint: you can show that this is indeed the case either by proving it directly or by exhibiting a counter example.*)
9. If x is an integer; find the solution set for the given problems, if none exists, write $\varnothing$. Write your results in set-interval notation whenever it seems most

appropriate to do so. (a) $x > 1$ and $x < 2$; (b) $x > 10$ and $x \geq 15$; (c) $y < 7$ or $y = 10$; (d) $x + 5 \geq 8$ and $x + 2 < 12$; (e) $x = 2$ or $x = 9$.

10. Verify that the extension to Theorem 2.2.1 is also true whenever $a < 0$ and $b < 0$. (Let the number of factors of a and b be 3.) ☆4 (see Appendix A for the solution to this problem)
11. Besides the visual difference between the comparative symbols used by axioms E5 and O4 (i.e., = and <), Axiom O4 requires the additional condition that $0 < c$, why? ☆5 (see Appendix A for the solution to this problem)
12. Prove that if $a < b$ and $c < d$, then $a + c < b + d$. *Hint: you will need the commutative property of addition (Axiom A2) to complete the proof (not to mention some axioms we have already covered).*

Chapter 3
The Axioms of Addition

What is addition? If it is only the act of joining something to something else, then we would have to admit that birds are phenomenal adders, building the incredible nests that they do one twig at a time. We would have to admit that countless other animals are phenomenal adders as well. From the American beaver to the corkwing wrasse, nature, it would seem, has endowed many animals with the ability to "add".

What most animals do not do, however, is keep count. In other words, they do not assign a unique sequence of grunts, whistles, or chirps for each new add-on (fortunately so, otherwise there would be no end to the incessant clamor of birds during nesting season). Those creatures that do keep count also remember what sequence of grunts, whistles or chirps follow any other sequence of grunts, whistles or chirps. Often too, these creatures will create pictures or symbols to represent each of these unique utterances.

By now I'm sure, you all know that I am speaking about *Homo sapiens*. However, if only humans can voice a unique utterance to determine the count, then it is very likely that neither Lucy of two million years ago,* nor the owner of the notched Ishango bone†, located at the Royal Belgian Institute (but not in its parking lot‡), could add beyond the count of their fingers and toes. Modern encounters with primitive societies suggest that they would have only been able to count, and therefore only add, up to three; beyond that primitive man would have made number quantity correspondences, for example, three chickens for every goat, or one pretty bead for one tasty fish.

It would seem then that some evidence of pictures or writing revealing an ability to count, as there can be no evidence of historical grunts, is required to determine not only what addition is but also who does or did addition.

By that reckoning, addition probably begins at around the time of the Mesopotamians some five thousand years ago, but let's not quibble, the Ancient Egyptians will do just fine—this is not, after all, a history book

The Ancient Egyptian method of addition was crude. They had a unique mark for tens of tens of things. The number one was a simple notch, ten was an inverted U, and a hundred looked like one of those spiraling all-day suckers, which may have represented the Pharos's crook or a coil of rope.

---

* The remains of a female skeleton of the hominin species discovered in 1974 in Ethiopia.

† A twenty-thousand-year-old, notched, femur bone, which is, supposedly, indicative of the first true inquiries into mathematics.

‡ A terrible joke.

Adding the number 223 to the number 118 would have looked something like the following:

Figure 8 Ancient Egyptian addition

In theory, it worked not too differently from our modern addition algorithm but without a place value system. In practice, especially when there were many numbers to add, it no doubt required considerably more writing space to perform, necessitating the use of "scratch paper", which in ancient times would have been rather expensive. It is more likely, therefore, that scribes and accountants would have used an abacus of sorts before recording their final calculation. Nevertheless, the Ancient Egyptians would have agreed with a few of the axioms of modern addition.

3.1 The Closure Property of Addition

One of the first properties we learned about numbers, although it was only implicitly taught, is that whenever we add a number to a number, we get another number. It is almost impossible to imagine otherwise. It is certainly true with the set of natural numbers: the sum of any two natural numbers is itself a natural number. However, not all things that start out as numbers, end up as numbers when they are added or multiplied, it all depends on how a culture defines numbers, and examples abound in the history of mathematics.

Some primitive tribes, right up until our modern era, had names for numbers only up to three, beyond that neither numbers nor addition existed. The Ancient Greeks, having once believed that all numbers were rational, were literally stupefied when they encountered irrational numbers. In later times, mathematicians had trouble accepting zero, later still the negative numbers, and the mother of all befuddlements, especially to young students, were the *imaginaries*: numbers that when multiplied by themselves yielded negatives.

To see how these troubles could arise, imagine that we restrict our definition of a number to be the statement *it is a number if and only if it is odd,* which happens to be the set $\{1, 3, 5, 7, \cdots\}$. If we add any number from this set to any other number from the set, say we add the numbers 139 and 47, both of which are odd, then a curious (or perhaps not so curious) thing happens, we obtain something that is not odd, we obtain instead the even number 186. However, from our statement, we must conclude that because it is not an *odd* number, it is not a number. In other words, we have obtained something that is outside of and alien to the set of numbers with which we started.

However, if instead we had restricted our definition of a number to be only the even numbers, which we will label $E = \{2, 4, 6, 8, \cdots\}$, then whenever we add any two numbers from this set, including the same number (e.g., $4+4$), we always arrive at another number already in the set. We call this set *closed with respect to ordinary addition*. Whereas the set of all odd numbers is *not closed* (to say it is *open* in this context would be poor form) with respect to ordinary addition. For future reference, we will denote the set of even numbers using ordinary addition as $[E,+]$ and the set of natural numbers using ordinary addition as $[N,+]$, both of which are closed under ordinary addition.

There are other types of sets and other operations besides addition (multiplication for example) that, depending upon context, may also be subject to the concept of set closure. However, before we get to those, we should have a more formal definition of the meaning of "set closure".

◦ DEF. 3.1.1: SET CLOSURE

A set is closed under an operation of mathematics, if whenever we apply the operation to any two elements from within the set, the result is another element in the set.

This definition can be used to construct or *generate* closed sets from sets that contain one, two, or any number of elements. For example, we can generate a closed set under addition from the set $\{0, 3, 5\}$, which is not closed, by including within it the sum of all possible pairings of the numbers in the set. Thus, we include the numbers $3+3$, $3+5$, and $5+5$, (0 has no effect) which yields the larger set $\{0, 3, 5\} \cup \{6, 8, 10\} = \{0, 3, 5, 6, 8, 10\}$. However, this set is still not closed, so we repeat the process and include the sum of all possible pairings of the numbers from this set, which are the sums (some of which are repeats)

$$3+6,\ 3+8,\ 3+10,\ 5+6,\ 5+8,\ 5+10,\ 6+6,\ 6+8,\ 6+10,\ 8+8,\ 8+10,\ 10+10$$

to obtain the even larger set

$$\{0, 3, 5\} \cup \{6, 8, 10\} \cup \{9, 11, 12, 13, 14, 15, 16, 18, 20\},$$

which is the equivalent set $\{0, 3, 5, 6, 8, 9, 10, 11, 12, 13, 14, 15, 16, 18, 20\}$. A quick check will reveal that this set isn't closed either, for example $11+18$ is not in the set, so we repeat the process. If we continue in this fashion indefinitely, we will have generated a very large (infinitely large) set for which the sum of any two numbers will ultimately be a number in the set; in other words, it will finally be a closed set (under ordinary addition).

There are also *finite sets* of numbers (sets with a finite number of elements) which are closed under addition, provided we don't mind fiddling with our idea of addition. For example, the set $\{\alpha, \beta\}$ is closed if we define $\alpha+\beta=\alpha$, $\beta+\alpha=\beta$, $\alpha+\alpha=\beta$, and $\beta+\beta=\alpha$. Although, this set will not satisfy the commutative axiom of addition, for the purposes of set closure that is irrelevant.

A more interesting finite set, which will satisfy the commutative axiom of addition, is the set defined by "clock" arithmetic. Consider Figure 9 (overleaf):

Figure 9

The set of numbers (hours) on a clock is $C=\{1, 2, 3, \cdots 12\}$. If we add any of these numbers to the time on the clock as hours, then using, what I will call, *wrap-around* addition, we will always arrive at some other number, or time, on the clock. For example, if we add 3 to the 10 on the clock, then we obtain 1, i.e., $10 \oplus 3 = 1$, which is a number in the set C.* If we add 10 to the 10 on the clock, we obtain 8, i.e., $10 \oplus 10 = 8$, which is, again, a number in the set. We could, of course, do this with any number in the set and any time on the clock and our additions in this manner would always give us a number that is in the set. We would have to try all the numbers to be sure, but it seems intuitively obvious. We will use the abbreviation $[C,\oplus]$ whenever we wish to talk about the set C of numbers on a clock along with its associated wrap-around addition.

However, we do not need the idea of wrap-around addition at all, we can choose to define all of the additions in a table, one-by-one, as was done in Table 4 (below) and examine the table, itself, for closure.

$\oplus$	1	2	3	4	5	6	7	8	9	10	11	12
1	2	3	4	5	6	7	8	9	10	11	12	1
2	3	4	5	6	7	8	9	10	11	12	1	2
3	4	5	6	7	8	9	10	11	12	1	2	3
4	5	6	7	8	9	10	11	12	1	2	3	4
5	6	7	8	9	10	11	12	1	2	3	4	5
6	7	8	9	10	11	12	1	2	3	4	5	6
7	8	9	10	11	12	1	2	3	4	5	6	7
8	9	10	11	12	1	2	3	4	5	6	7	8
9	10	11	12	1	2	3	4	5	6	7	8	9
10	11	12	1	2	3	4	5	6	7	8	9	10
11	12	1	2	3	4	5	6	7	8	9	10	11
12	1	2	3	4	5	6	7	8	9	10	11	12

Table 4: Wrap-around addition table for the number 12

It is plain to see that $[C,\oplus]$ is closed under this definition of addition.

There is one more (contrived) set that I would like for us to consider before we get to the first *proper* axiom† of algebra (but otherwise the eleventh by our count), but don't worry, it won't be around for long. The elements of this set

---

* We use symbols such as $\oplus$ and $\otimes$ to denote a type of addition or multiplication which is different from our ordinary understanding of addition and multiplication.

† The closure axioms are rarely discussed with or even presented to high school students, which is most unfortunate, not only because it is an easy enough concept to understand, but because it is an important concept throughout much of "higher" mathematics, most notably Linear and Abstract Algebra.

are, again, all of the natural numbers, which we will denote, as usual, by $N = \{1, 2, 3, \cdots\}$. However, this time addition under the set of natural numbers will be defined rather oddly; hence, we will use the symbol "♣" (read as "clubs") to denote this type of addition. It goes something like the following:

If a and b are elements of N, then $a ♣ b = 2(a+b)$. ▪3.1.1

Do not let the definition of ♣, which involves both the ordinary addition of two numbers and the multiplication of a number (by 2), confuse you. The word *addition* is just a word, it can mean whatever we want it to mean. What matters is whether this definition of addition will conform to and meet the requirements of the axioms.

Let's try an example to get a feel for how this addition works: 6 and 11 are certainly elements of N, then, according to the definition presented in 3.1.1, to "add" the two numbers we first add the numbers in the usual fashion, get the result, and then multiply it by 2. Hence, $6 ♣ 11 = 2(6+11) = 34$. (The positions of the numbers must be kept in the order indicated.)

It should be clear that no matter what pair of natural numbers we choose to add in this fashion, we get another natural number, albeit a number which, in this case, is always even, but a natural number nevertheless. Try a few on your own if you need more convincing that addition defined in this queer manner on the set of natural numbers will always be a natural number and that, therefore, the set is closed under this definition of addition. We will denote this set of natural numbers, N, along with its associated addition, ♣, as $[N, ♣]$.

Whenever we talk about a set along with one or more operations, addition, multiplication, or what have you, we are talking about *systems*. Thus, $[E,+]$, $[N,+]$, $[C,\oplus]$, and $[N, ♣]$ are called systems and we will examine these four systems in relation to the axioms throughout the course of the book.

With that, we come to the first, proper axiom of algebra. It states that the set of numbers with which algebra is concerned *must* be a closed set under addition. Called the closure property of addition, it reads as follows:

❖ Axiom A1: Closure under Addition

If a and b are (any) numbers, then $a+b$ *is a number*

Because of the symmetrical property of the English word *and* in the above axiom and because the axiom is a statement about numbers in general, it goes without saying that if a and b are numbers, then b and a are numbers (you cannot argue with that); consequently, the compound symbol $b+a$ is also a number (although not, at least for the time being, logically equal to $a+b$).

As you can see from Figure 10 (next page), two numbers go into the addition machine according to Axiom A1, but only one number comes out, and whichever number goes into the addition machine first determines how the machine performs the addition.

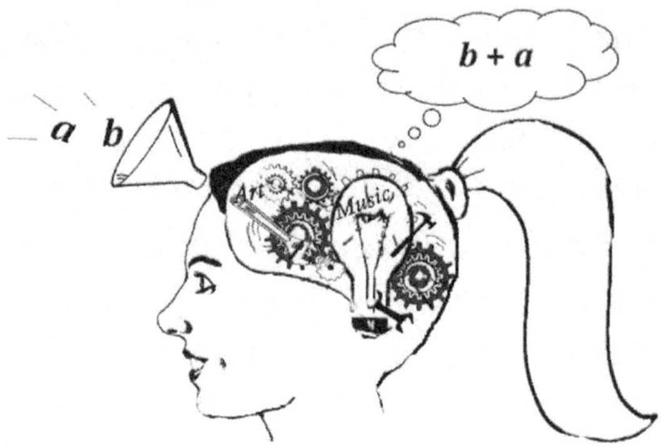

Figure 10: Jessi (the adding machine) adds a to b.

An important aspect about set closure is that if we can show that a set is closed (and to do that we only need to show that the addition [or multiplication] of any two numbers from the set is a number in the set), then the addition of any three numbers from the set is a number in the set, and the addition of any four numbers is in the set, and so on. The reason is that the addition of any three, four, or more numbers is, in many ways, the addition of only two numbers—it is a matter of perception. The wrap-around addition table above bears this out quite well. Try adding three or more numbers from the clock and you will always be adding exactly two numbers at a time. More on this when we get to the commutative and associative axioms of addition.

Just as important as what this axiom says, is what it doesn't say. It doesn't say that a and b cannot be the same number, only that they are numbers; thus $a+a$, $b+b$, etc. are also numbers. It also doesn't say that given three (or more) numbers, a, b and c, it follows that $a+b+c$ (with no parentheses) is a number (although later we will prove it is). If it were to say that instead of what it does say, then we would not only be at a complete loss describing what the sum of two numbers could possibly be (a salted anchovy perhaps?), but, what is more, we would also never be allowed to add two numbers at a time, precisely because the addition of two numbers would not have been defined.*

Thus, the axiom *must* say that if we are given a *pair* of numbers, then the addition of the given pair will itself be a number; but that has its drawbacks as well. We are restricted to working with just the numbers that we are given, in this case just a and b, and the pairs (and the pairs of pairs and the nesting pairs) of numbers we can construct from them.

Therefore, if we are given any *three* (distinct) numbers, a, b and c, then the following twenty-one items would also be numbers according to Axiom A1. Of course, as was mentioned earlier, there would be an infinitude of numbers constructible from these three numbers, which would mean that numbers like $((a+a)+a)$, $((((b+b)+b)+c)+c)$, etc., would also need to be included, but we

* Kids understand this logic implicitly; one can hear them saying "you didn't say it couldn't be the same number" or "you didn't say what two is, you only said what three is".

will restrict our discussion to just the following twenty-one numbers:

$$a, b, c, a+b, b+a, a+c, c+a, b+c, c+b,$$
$$c+(a+b), c+(b+a), b+(a+c), b+(c+a), a+(b+c), a+(c+b)$$
$$(a+b)+c, (b+a)+c, (a+c)+b, (c+a)+b, (b+c)+a, (c+b)+a. \qquad \blacksquare 3.1.2$$

The parentheses, which may not be removed (yet), are needed to indicate the pairwise additions of these three numbers (i.e., a, b and c).

However, this axiom says nothing that would allow us to place an equality symbol between any pair of the above twenty-one numbers and make the claim that they are equal. For now, all that we can say for certain is that the twenty-one objects in line 3.1.2 are numbers (in a logical, pseudo-number sense) — that is the extent of the axiom.

It also needs to be pointed out that because this axiom does not (nor does any axiom) assign a definition to the word "number", it is open to interpretation. If we think of a number in the usual sense, then this axiom limits our choices of what we may consider a suitable set of numbers for algebra. We have already shown that the set of odd numbers is not closed under addition. Hence, algebra cannot be just about the odd numbers. On the other hand, the set of even numbers, $E = \{2, 4, 6, \cdots\}$, and the set of natural numbers, $N = \{1, 2, 3, \cdots\}$, are both closed under ordinary addition (and ordinary multiplication for that matter), which we have previously represented as $[E,+]$ and $[N,+]$ respectively. Hence, in the absence of any other axioms that might eliminate either of these two systems from algebra, they are both acceptable systems to use with the axioms of algebra.

Because there is no assigned definition to the symbol "+" (plus/addition) either, algebra could also be about the twelve natural numbers on a clock along with its associated wrap-around addition: $[C,\oplus]$. It could also be about the set of natural numbers along with the strangely defined addition, $\clubsuit$.

Any of these four sets of numbers along with their associated methods for adding them (i.e., the systems $[E,+]$, $[N,+]$, $[C,\oplus]$ and $[N,\clubsuit]$) may be acceptable to use with the axioms of algebra. However, they must meet the demands of *every* axiom, otherwise they must be excluded from consideration.

It is possible, though, that by extending our definition of a number to any of these sets to include other numbers, which we may encounter along the way, we may be able to expand each set of numbers so that each meets the demands of those axioms, which would otherwise exclude them, at least until such time that it becomes unreasonable to try to salvage them any further.

However, a word of caution is in order. Whenever a change is made to a system, either by including more numbers in the set or by redefining the addition, or even by adopting other operations (multiplication for example), then it must be rechecked for conformity to every axiom all over again.

Besides limiting what we may consider as a usable set of numbers for algebra, there is another important characteristic of this axiom. It lies in the fact that it makes a logical statement about all things known as numbers. Therefore, whatever can be logically said about a "single" number, can be logically said about the sum of any two numbers, because the sum of any two numbers is, by this axiom, a (single) number. In other words, any axiom, theorem or definition, which can be applied to the numbers a and b individually, can be applied to the number $a+b$ as well. Furthermore, whatever can be said about the number $a+b$, can also be said about the number $(a+b)+(a+b)$ and the number $(a+b)+b$ and so on, because the closure property guarantees us that all of these sums are also numbers in the set. It's like moving a mountain with no more effort than the force required to push a pebble.

This manner of constructing the many varied numbers in a set is a result of the self-referential or recursive feature of this axiom. Hence, if c is any number in the set, then $(((c+c)+c)+c)+c$ is also in the set. Having axioms refer to themselves is a remarkable idea, which we will encounter again.

From an interpretive point of view, this axiom—along with the closure axiom of multiplication—implies that no matter how complicated a number expression may appear, it is, or it always points to, or more accurately it is always equal to exactly one number in the set (there are no repeats). For example, whenever we see a somewhat complicated group of symbols as in the expression $2\cdot(17+34)+1$, we recognize it as being equal to (a repeat of) the abstract number one hundred-three. By analogy, the compound symbol $a\cdot(b+c)+d$ (the form of which is identical to the numerical example just given) is also logically equal to exactly one number in the set and consequently it may be treated as though it is one number—that is what the closure axioms do.

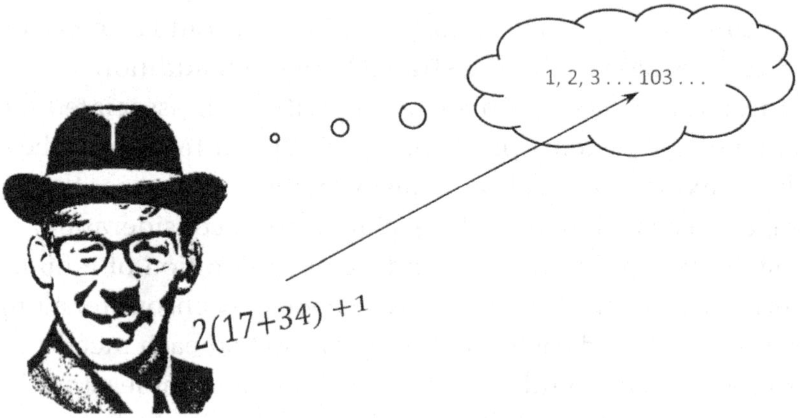

Figure 11: Tony contemplates the uniqueness of numbers.

The closure properties also lend substance to the axioms of equality and inequality, which, if you recall, presented to us the symbol "+", but gave no indication about what it was, from where it came or how to get one. Consequently, (e.g.) Axiom E4, which stated, *if a, b, and c are numbers and* $a=b$, *then* $a+c=b+c$, is now comprehensible as equality between two *numbers*.

3.2 The Commutative Property of Addition

The order in which we add one number to another number will not change the sum. Thus 4 added to 3 is the same number as 3 added to 4. We know this about numbers from arithmetic. We can show it pictorially as,

$3+4$

$4+3$

Symbolically, we can say that the order in which we write the numerals on either side of the plus symbol does not matter.

This property of arithmetic is called the *commutative property of addition* because the numbers commute back and forth. You can remember it by thinking about how you commute from home to school and back again, or to work and back again as may be the case.

❖ **Axiom A2: Commutativity of Addition**

If a *and* b *are numbers, then* $a+b=b+a$

Axiom A1 assures us that both $a+b$ and $b+a$ are numbers, otherwise Axiom A2 would make no logical sense. (We would not know what $a+b$ and $b+a$ represent.) Furthermore, because this axiom utilizes the equality symbol, we may, of course, use all of the axioms associated with equality, as we see fit.

However, this axiom only allows us to claim equality between re-ordered pairs of numbers. Referring to the collection of twenty-one numbers in line 3.1.2 of the previous section, we can now say that of those numbers we may claim that $a+b=b+a$, $a+c=c+a$, and $b+c=c+b$.

Making use of this axiom and the substitution axiom of equality (E6), we may also partition twelve of the numbers from line 3.1.2 into three distinct groups of equivalent numbers, but only in the distinct groups shown in Figure 12. (The "teepees" in the figure indicate pairs of, and nesting pairs of numbers.)

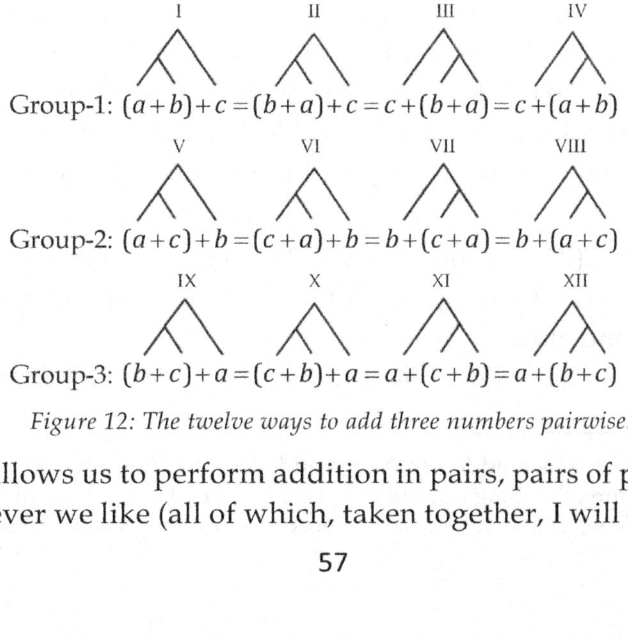

Figure 12: The twelve ways to add three numbers pairwise.

Axiom A1 allows us to perform addition in pairs, pairs of pairs, and in nesting pairs however we like (all of which, taken together, I will call parenthetical pairs), which, for three numbers, accounts for the twelve distinct numbers shown in Figure 12. Axiom A2 allows us to determine which of these twelve numbers are equal, but only if there has only been an order reversal of pairs. It does not allow us to cross over to distinctly different pairs and make the claim that they are equal. We cannot, for example, say that the number $(a+b)+c$ from Group-1 is equal to the number $a+(b+c)$ from Group-3, or that either of these two numbers are equal to the number $b+(a+c)$ from Group-2 because each of these numbers have distinctly different pairs, and reversing the order of these pairs will not change that distinction.

It should also be clear that the way we perform the addition of the number $(a+b)+c$ is different than the way we perform the addition of the number $a+(b+c)$. The first expression reads, "add c to the sum of b added to a"; while the second expression reads, "take the sum of c added to b and add it to a".*

Without a rule that would allow us to claim equality between the numbers from different groups, as in $(a+b)+c$ from Group-1 and $a+(b+c)$ from Group-3, then a simple statement such as $(1+2)+3=1+(2+3)$ can never be logically obtained, and the three groups would necessarily stand apart from one another. On the other hand, if we had such a rule, then we would be able to effectively change the pairs from one group into the pairs of another group and make them all equal. What we need then is another axiom.

However, before we turn our attention to the associative property of addition, which will do just that, there is the small matter of determining whether our four systems, $[E,+]$, $[N,+]$, $[C,\oplus]$ and $[N,\clubsuit]$, are in fact commutative. We will not prove that they are, but only let our good intuitions guide us.

In the case of $[E,+]$, it should be apparent that if we add (using ordinary addition) two even numbers in any order the result will remain unchanged: $2+6=6+2$. The same is true of $[N,+]$. If we add two natural numbers in either order, the result is unchanged. In the case of $[C,\oplus]$, it should also be apparent that whether we add 8 hours onto 10:00 o'clock or 10 hours onto 8:00 o'clock the result will be 6:00 o'clock. Thus, $10\oplus 8=8\oplus 10$, and any pair of clock numbers will commute with equal results, Table 4 also bears this out. For $[N,\clubsuit]$, it is probably a good idea to try a couple of examples to get a sense of whether it is commutative.

- *Ex.1*: $21 \clubsuit 9=2(21+9)=60$ and $9 \clubsuit 21=2(9+21)=60$, so $21 \clubsuit 9=9 \clubsuit 21$. That checks out. Although, it should be clear that $[N,\clubsuit]$ is commutative, we will try one more.
- *Ex.2*: $191 \clubsuit 8=2(191+8)=398$ and $8 \clubsuit 191=2(8+191)=398$. *Check.*

---

* Whether we view $a+b$ as a added to b or b added to a is not important, provided we are consistent. Furthermore, detailing how addition is performed helps students to see the purpose of the associative (the next) axiom and the purpose of parentheses in general.

Although this is only a verification, it should be clear that changing the order of the numbers on either side of ♣ will not affect the result of that addition, because whether ♣ is commutative is dependent upon whether ordinary addition is commutative, which it is; hence, so is ♣.

As it stands, all four of these systems are still eligible to be used as objects for the axioms.

~

3.3 The Associative Property of Addition

How we group numbers for addition will not change the sum. Taking the sum of 4 added to 3 and adding that to 2 is the same as adding 4 to the sum of 3 added to 2. We know this about numbers from arithmetic. We can show it pictorially:

$2+(3+4)$ $\qquad$ $(2+3)+4$

Using alpha-numbers to represent the equality between these two distinct methods for doing the addition, we could write, $a+(b+c)=(a+b)+c$. Making use of the symmetric property of equality (E2), we may also write, $(a+b)+c=a+(b+c)$, which brings us to the next axiom.

❖ **Axiom A3: Associativity of Addition**
If a, b and c are numbers, then $a+(b+c)=(a+b)+c$

You can remember the name of this axiom by thinking about which numbers are "associating", via parentheses, with one another.

Recall, from Section 3.1, that the closure axiom of addition allowed us to perform the addition of three numbers in twelve ways, none of which could be said to be equal based on that axiom alone. The commutative axiom stated that of those twelve whose pairs were only reversed are equal. That left us with three groups of numbers (Figure 12 on page 57) where the numbers in any one group were not equal to the numbers in any other group. With this axiom, however, we can now prove that all of the numbers are equal.

Proof: Consider again Figure 12, Axiom A3 states that $a+(b+c)=(a+b)+c$. (These are the numbers labeled "XII" from Group-3 and "I" from Group-1, respectively.) Hence, every number in Group-3 is now equal to every number in Group-1 by the repeated use of the transitive property of equality (E3). However, applying Axiom A2 we have $a+(b+c)=a+(c+b)$, and because Axiom A3 is a general statement about *all* numbers, we can "shift" the parentheses* on the right hand side of this equality and write $a+(c+b)=(a+c)+b$. Invoking the transitive property of equality (E3), we are able to conclude that $a+(b+c)=a+(c+b)=(a+c)+b$. However, $a+(b+c)$ is a number from Group-3 and the

* Technically, we are not "shifting" parentheses, but it is convenient to think as though we are.

number $(a+c)+b$ is from Group-2. Thus, like a perpetual waterfall, every number in Group-2 is equal to every number in Group-3, and every number in Group-3 is equal to every number in Group-1. Ergo, they are all equal.

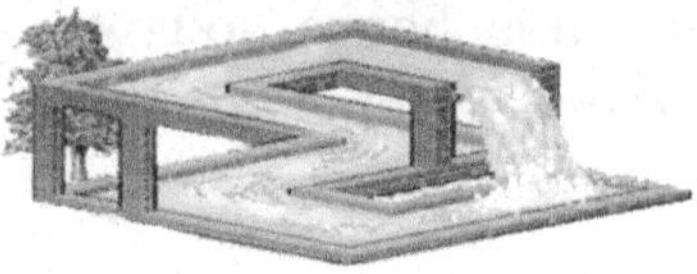

The importance of this brief discussion about the commutative and associative properties is not just to explain the logical consequences of these axioms, although it is primarily that, but also to inform us of the many ways in which we can perform addition and the many ways in which we can communicate our thoughts about addition.

For example, there are two ways to add two numbers, twelve ways to add three numbers* (as already shown), 1,680 ways to add five numbers, and, if you need a little more perspective on the matter, then consider this: there are close to twenty billion ways to add ten numbers; all of them are written slightly different from one another—all of them communicate a slightly different thought—yet they are all equal! (For those of you who find this count impossible to believe, then I urge you to find the exact count, ☆6—*hint: find a pattern.*)

There is another reason why we do not add more than two numbers at a time besides the reason given in Section 3.1 (where we explained that it was a matter of logical necessity): we are simply incapable of adding more than two numbers at a time. It is not a matter of choice, rather it is a limiting characteristic of the human mind—we are only capable of adding two numbers at a time. We obtain the sum of any two numbers and then add another number to that sum to obtain yet another sum, then add another number to that sum, and so on, and we continue in this fashion of adding pairs until all the numbers have been added. Even when we think we are adding three numbers at a time, as in the addition problem in Figure 13, we are still only ever adding two numbers at a time.†

$$\begin{array}{r} {\scriptstyle 1} \\ 11 \\ 35 \\ +\ \ 7 \\ \hline 53 \end{array}$$

Figure 13: Elementary addition is always done with pairs of numbers.

It should be clear upon carful reflection that every step of the addition

* Elementary school students should be encouraged to discover the many ways in which two, three or even four number can be added and multiplied in parenthetical pairs.

† This fact is rarely brought to the direct attention of children in the elementary grades. Yet reflection on the actual mental process of addition—two numbers at a time—will not only help them with the performance of addition, but it will also prepare them for algebra.

procedure in Figure 13 never involves more than two numbers at a time. We add 5 to 1 to get 6, we add 7 to 6 to get 13, and so on. If there are any skeptics among my readers, then I challenge you to add the numbers 9, 3, 8 and 7 at once. Not in pairs mind you, but all together, at once.

Perhaps in some future world, humankind may have evolved to be able to mentally add more than two numbers at a time; perhaps a secondary brain will have developed from some large bulbous mass just above the central forehead giving rise to the faculties of a "duel-core" brain. Who can say, the future can be a billion years and beyond from now.

In the meantime, it is difficult to imagine how we would even communicate such an idea mathematically, never mind actually performing it, but Figure 14 is an attempt, feeble as it may be, to do just that. (One wonders what the properties of this arithmetic would be.)

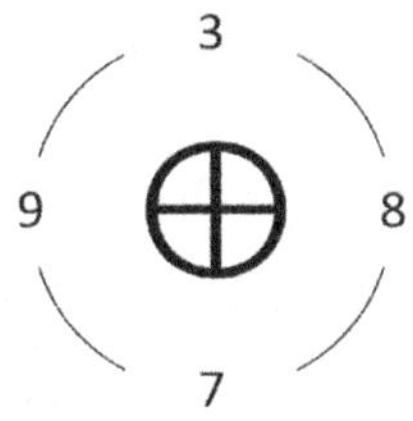

Figure 14: Duel-core-brain addition.

Addition, like the union and intersection of two sets in Section 3.1, is considered a binary operation, which would make duel-core-brain addition a sort of quaternary operation.

Of course, just because we cannot add more than two numbers at a time, it doesn't necessarily mean that the associative axiom of addition couldn't have been written to include more than associating two numbers at a time; so why doesn't it? The answer is that there is no need for it to. We can demonstrate that grouping two numbers at a time implies the grouping of three, four or more numbers at a time.

Consider the five possible parenthetical pairings of the four numbers, a, b, c, and d, (the arrangement of which remains unchanged) as shown in 3.3.1.

$$((a+b)+c)+d=(a+(b+c))+d=a+((b+c)+d) = a+(b+(c+d))=(a+b)+(c+d). \quad \blacksquare 3.3.1$$

From the first expression to the last, Axiom A3 has been applied four times in succession. Observing how the parentheses move through each consecutive expression, you may have noticed that it has a pattern suggestive of an inchworm's motion. That's because Axiom A3 restricts the shifting of parentheses (in this case to the right) between pairs of numbers in such a way that the outer most parentheses cannot move to the right until all of the inner parentheses have moved to the right. In other words, depending on the initial position of the parentheses, there may be no way to logically arrive at another position of the parentheses except through some intermediary step (or steps), hence the inchworm effect. For example, there is no logical way, given the pairwise restrictions of Axiom A3, to get directly from $a+((b+c)+d)$ to $(a+b)+(c+d)$ without the intermediate step of either $a+(b+(c+d))$ or $((a+b)+c)+d$.

It is important to point out, however, how we got to the last expression in 3.3.1 from the one immediately before it. Specifically, we need to point out how we arrived at the equality

$$a+(b+(c+d))=(a+b)+(c+d).$$

The reason (if you recall) is that $c+d$ may be thought of as a single number (by the closure property) in the same way that $7+2$ may be thought of as the single number 9. All numbers, which are enclosed within parentheses (or even if they are not, depending on how we wish to interpret an expression), may be considered a single number. With that said, if we were to replace $(c+d)$ with some single symbol to represent it, say $m=(c+d)$, then in terms of form it would resemble (shown underlined below) the associative property precisely:

$$a+(b+(c+d))=\underline{a+(b+m)=(a+b)+m}=(a+b)+(c+d).$$

It should be apparent from the string of equalities in 3.3.1 (previous page) that we have accounted for every possible grouping by pairs and nested pairs of these four numbers a, b, c and d, for the arrangement shown, and Axiom A3 implies that they are all equal. Therefore, we can remove the parentheses altogether and add the numbers in pairs however we choose, provided they are adjacent pairs and we don't change the arrangement because doing so would require the commutative property (and we are strictly speaking about the associative property at the moment). In other words, we may write the sum $a+b+c+d$ without any parentheses at all or, for that matter, place parentheses wherever we like, as in $(a+b+c)+d$, without affecting the sum. Thus, we have shown that Axiom A3 only needs to consider groupings of numbers in pairs and that all other groupings will consequently follow.

Before we continue, a further word about parentheses is in order. Parentheses "( )”, square brackets, “[ ]” or occasionally curly braces, “{ }”, are used to communicate precisely how we wish to identify numbers. So, in addition to grouping numbers in pairs, we can also communicate the idea of adding *four* numbers (Each number is called a *term* whenever it is separated from other numbers by a plus symbol.) as in the following expression, which contains no parentheses (and four terms):

$$\overset{1}{a}+\overset{2}{b}+\overset{3}{c}+\overset{4}{d};$$

or we can also communicate the idea of adding *three* numbers (three terms) as in any of the following three expressions at the top of the opposing page:

$$(a\overset{1}{+}b)\overset{2}{+}c\overset{3}{+}d \text{ or } a\overset{1}{+}(b\overset{2}{+}c)\overset{3}{+}d \text{ or } a\overset{1}{+}b\overset{2}{+}(c\overset{3}{+}d);$$

or we can communicate the idea of adding *two* numbers (two terms) as in any of these three expressions:

$$(a\overset{1}{+}b+c)\overset{2}{+}d \text{ or } a\overset{1}{+}(b+c\overset{2}{+}d) \text{ or } (a\overset{1}{+}b)+(c\overset{2}{+}d);$$

or we can communicate the idea of the single number (one term)

$$(a+b+c+d).$$

If we include the five associative groupings presented in line 3.3.1 with the seven new groupings presented above,* then there are twelve ways in which we can communicate our thoughts about *grouping* the four numbers $a, b, c,$ and d using the language of mathematics (for just this arrangement).

It also needs to be said that when we use parentheses, we must do so in such a way that each opposing pair of parentheses (i.e., "( )") is either nested within another opposing pair or alongside such a pair. Pairs of parentheses may not overlap. For example, $(a+[b)+c+d]$ is ill formed because the pairs of parentheses overlap; consequently, how the additions are to be carried out is garbled and nonsensical (and impossible to produce according to Axiom A1). On the other hand, both $a+[(b+c)+d]$ and $(a+b)+(c+d)$ are well formed because the pairs of parentheses are either nested within or alongside one another, and it is clearly understood how the additions are to be carried out.

Before we move onto the next axiom, there is the question of whether our four candidate systems satisfy the conditions of the associative axiom of addition. In the case of $[E,+]$, using the three even numbers 8, 22, and 88 we have

$$8+(22+88)=108 \text{ and } (8+22)+88=108. \textit{ Check.}$$

Clearly, the associative axiom has no effect on $[E,+]$. It should also be clear that $[N,+]$ also satisfies the conditions of the associative axiom. I leave it to you to check the accuracy of that statement.

What about clock arithmetic, $[C,\oplus]$? Does, say, $9\oplus(11\oplus 6)=(9\oplus 11)\oplus 6$?

$$9\oplus(11\oplus 6)=9\oplus 5=2 \text{ and } (9\oplus 11)\oplus 6=8\oplus 6=2. \textit{ Check.}$$

Without proving that $[C,\oplus]$ is associative and without checking every pairwise addition, it seems reasonable based on the wrap-around addition table (Section 3.1) and a smattering of mathematical intuition that $[C,\oplus]$ is indeed associative.

Last, we come to $[N,\clubsuit]$. Does $9 \clubsuit (11 \clubsuit 6)=(9 \clubsuit 11) \clubsuit 6$? We put it to test:

$$9 \clubsuit (11 \clubsuit 6)=9 \clubsuit [2(11+6)]=9 \clubsuit [2\times 17]=9 \clubsuit 34=2(9+34)=2\times 43=\underline{86}.$$

(When doing this type of addition, it is important to follow the pattern.) Hence, $9 \clubsuit (11 \clubsuit 6)=86$, whereas

$$(9 \clubsuit 11) \clubsuit 6=[2(9+11)] \clubsuit 6=[2\times 20] \clubsuit 6=40 \clubsuit 6=2(40+6)=2\times 46=\underline{92}.$$

---

* *Note*: the grouping $(a+b)+(c+d)$ appears twice: once in 3.3.1 and again above.

Hence, $(9 \clubsuit 11) \clubsuit 6 = 92$. That does *not* check! $86 \neq 92$. It only takes a single *counter example* to disprove any general statement* and here we have shown that $9 \clubsuit (11 \clubsuit 6) \neq (9 \clubsuit 11) \clubsuit 6$. It seems highly unlikely that introducing new numbers could make $[N,\clubsuit]$ associative either because the problem seems to lie not with the numbers but with the way we have defined the addition. Without any clear way to make $[N,\clubsuit]$ associative, short of redefining the addition itself, which defeats the purpose of having $\clubsuit$ in the first place, we must exclude $[N,\clubsuit]$ from the running. We are now left with just the three candidates: $[E,+]$, $[N,+]$, and $[C,\oplus]$.

~

3.4 The Additive Identity Element, 0

We now come to the number that many civilizations over many centuries have tossed out as either ridiculous or downright scary: the number zero. Those ancient and later civilizations that did use it, understood it in a completely different way than how we understand it today. It was strictly used as a place holder to indicate the absence of an existent number.

Consider the number 1,024 for example. To understand this number, we need to understand the meaning of each position in the number. The number "4" on the far right means that there are four ones. We are not multiplying, mind you, we are counting how many ones we have, and we have four of them. The number 2 to the left of that means that there are two tens to add to the count. To the left of that is a placeholder, 0, indicating that we do not have any hundreds to add, so we move on to the next digit, which is a 1, indicating that we have one thousand to add. Putting it all together we have

$$1{,}000+10+10+1+1+1+1=1{,}024.$$

Of course, the symbol 0 cannot be removed from the above number without changing the value of the number; its position in the number is essential. It signifies that we are not adding any hundreds.

However, not adding something is a completely different paradigm than adding nothing! We don't add things all the time. In the number 1024 we didn't add any hundreds. In the expression $2+3$, the number 27 wasn't added. Not adding things was not a problem for past civilizations either: not adding the neighbor's goats to one's own goats was an every-day occurrence. Civilizations have been *not adding things* since the dawn of mathematics, but *adding nothing*, and creating a symbol for this "nothing", which is the thing to be added to something, required an intellectual leap that we can scarcely now imagine, especially when coupled with the fear that many people, of mostly Western civilizations, had about the dark, eternal "Zero" of the underworld that awaited

* Thus, the claim "all turtles are toothless" is disproved if a single toothed turtle is found.

a wicked and immoral life. If nothing else, it attests to the power of the mind and mettle of the spirit to venture into darkness and overcome the unknown.

However, over the course of some centuries, probably around the fifth century in India (earlier in Mayan civilizations), humans had finally decided and declared that the number 0 exists as a number. (It provided an answer to the question, "what number when added to one is one?")

To this day, however, the philosophical problems surrounding zero still emerge from time to time. Most notably is that if zero is nothing, then it is in some respect infinite, but at the other end of the infinity that is at the edge of everything. (A thought that troubled many bygone mathematicians.)

The number zero is quite possibly the point in all of space and time, and human history when mathematics either diverged from the real world or converged to meet it head on. Perhaps, we will know which of the two it is when, if ever, we finally understand the cosmological objects known as Black Holes.

We are all familiar with most of the ways that 0 behaves. First and foremost, of course, is that when we add it to a number, or a number to it, the result of the addition is the same as the number itself. Hence, $7+0=7$ and if a is a number, then $a+0=a$. One could say that it returns the *identity* of the number to which it is added. Anything that possesses this property is a zero. It is not important what zero looks like, only that it behaves like zero. Well, without further ado, allow me to introduce Axiom A4, the additive identity axiom.

❖ **AXIOM A4: THE EXISTENCE OF THE ADDITIVE IDENTITY NUMBER**

There exists a number called zero, which is written as 0*, such that if* a *is a number, then* $a+0=a$

The single most important thing to notice about this axiom is that it clearly states there *exists* at least one thing that is an "authentic"* algebraic number, and that is the number 0. Every other axiom that we have encountered to this point only presumed the existence of numbers by way of using the conditional statement "*if* such and such are numbers, then ..." By contrast, this axiom comes right out and makes the bold claim that 0 *is* a number (no *if*s about it) and that it has the above stated properties. Apparently then, there is a set of numbers yet to be discovered and constructed that conforms to *all* of the axioms of algebra and which contains, minimally, the number 0.

However, because 0 *is* a number, the closure axiom states that every addition of 0 with itself, i.e., $0+0$, $(0+0)+0$, $(0+0)+(0+0)$, $((0+0)+0)+0)$, etc., must also be numbers. Consequently, we have at a minimum the following "legal" set of numbers which conforms to all the axioms of algebra:

$$\{0,\ 0+0,\ (0+0)+0,\ (0+0)+(0+0),\ ((0+0)+0)+0)\ \cdots\ \}. \qquad ▪3.4.1$$

Because 0 and + are just symbols, we would normally have no way of

* "Authentic" will mean, *in this book*, those numbers generated by the axioms themselves.

knowing what $0+0$ equals. However, this axiom states that if we "add" 0 to any number—that is if we put a plus symbol between it and any number—it returns the identity of the number, which, in this case, is none other than the number 0 itself. Hence, $0+0=0$. Which number is playing the role of zero and which is playing the role of the number whose identity is being returned is your pick because of the obvious fact that they look identical. Employing the substitution axiom of equality (E6), it follows that every pair of 0s and every pair of pairs of 0s, and every nested pair of 0s in line 3.4.1 is just the number 0. Thus, for example, $(0+0)+(0+0)=0+0=0$. And like a string of falling dominoes, the set consists of nothing but the single element 0.

We will call this set the discovery set, D, for the moment and write $D=\{0\}$. It currently only contains the element zero, but it could possibly contain other (authentic) numeral-numbers which have yet to be identified.

Because 0 *is* a number,* there is no need to determine if it satisfies the conditions of axioms A2 and A3 (the commutative and associative axioms). It is a number by axiom and therefore it satisfies the conditions of every axiom of algebra. This was not the case with every other set of numbers we have examined so far (i.e., the sets E, N, and C, of the systems $[E,+]$, $[N,+]$, $[C,\oplus]$ and $[N,\clubsuit]$). In each of those cases, "number-hood" was granted only if the set could meet the conditions of the axioms—it was not granted carte blanch. For example, it may be customary to call things, like 2, 11 and 99 numbers, but that does not make it so from the perspective of algebra. The set of numbers {3, 5} for example, is not a usable set of numbers for algebra because, first and foremost, it does not meet the first condition of number-hood: the set is not closed. Hence, 3 and 5, by their lonesome, are not, *logically* speaking, numbers.

The last thing to be said about the axiom-generated, numeral-number zero is that it is unique (so is every other number, but mathematicians like to pick on zero for some reason); that is to say, 0 not only exists, but it is the only such number that exists with the property of zero. For suppose there is another numeral-number, say O, that has the same properties that 0 has, then, because both 0 and O are numbers, it follows that $0+\mathrm{O}$ is a number (A1). Because they both return the identity of any number added to them (in this case only by assumption for O), it follows that $0+\mathrm{O}=0$ and $0+\mathrm{O}=\mathrm{O}$; then by the transitive property of equality we have that $0=\mathrm{O}$. In other words, we only need the numeral-number 0 to do the job of zero.†

Considering all of the above, it is more than appropriate to include $[D,+]$ as a shoo-in for the game of algebra, which, for now, is the set consisting of the

* It is important that I stress this logical fact.

† Think of it this way: suppose there exists a comet "out there" with Earth's name on it. That statement does not preclude the possibility of there being more than one such comet. Of course, it would certainly be much better for life on earth if such a comet was unique.

single element 0 under "+" (addition). However, this set will grow infinitely large because, as we shall see, 1 is also a number given to us by the axioms, and because it *is* a number, the closure axioms will require us to determine the infinitude of numbers constructible from it.*

Because the set D arose from the axioms, it is special compared to the sets of the other systems. Hence, to remind us of its special place among the contenders, we will attach a small subscript "A" (for Axiom) to the system notation, thus we will use the notation $[D,+]_A$ to represent the set consisting of the single element 0 (for now) under "+" given to us by the axioms.

It is significant to observe that everything we have just discussed about zero, had absolutely nothing to do with anyone's *beliefs* about the number zero or about addition. That is to say, $0+0$ equals 0 not because of some steadfast belief that nothing added to nothing is nothing, but because the symbol 0 has the *logical* property outlined in Axiom A4.

Before we move onto the next axiom, there is the question of whether $[E,+]$, $[N,+]$, and $[C,\oplus]$, are still viable systems for algebra. (We know that $[D,+]_A$ is by axiom.) With respect to Axiom A4, the question is whether each set contains an identity element (i.e., a zero).

In the case of $[E,+]$ then, we need to decide whether there is a number in the set $E=\{2, 4, 6, \cdots\}$ that behaves like zero. The short answer is a flat no. It is plain to see that every even number when added to another even number will produce an even number that is *greater* than either of the two numbers we started with and not, therefore, identical to either one of them, which is the behavior we seek. Hence, $[E,+]$ does not have a zero element. However, if we call the number 0 an even number and include it in the set E and label the new set as $E_0=\{0, 2, 4, \cdots\}$, then this set does have the zero element. However, because the set has changed, we must go back and check to see that this new set, E_0 (along with its ordinary addition), still conforms to the previous three axioms. I will leave it to you to convince yourself that indeed it does. We will label this new, revised candidate system as $[E_0,+]$.

In the case of $[N,+]$, where $N=\{1, 2, 3, \cdots\}$, it is also clear, for the same reasons as with the set E, that there is no zero element. However, we can add it to the set N as well. We will then have the revised set $N_0=\{0, 1, 2, \cdots\}$, which does have the zero element. However, at the risk of repeating myself, we must check that our new addition to the set did not disrupt the conformity of this system to the axioms, and we must, therefore, check it again against all the previous axioms. We will label the new system $[N_0,+]$.

Lastly, with respect to the clock system, $[C,\oplus]$, it turns out that the set $C=\{1, 2, \cdots 12\}$ of clock numbers along with wrap-around addition does have a

* The axioms state that these two numbers, zero and one, and all numbers derivable from them, like $1+1$, $1+1+1$, etc., are already in the set D, we just have to figure out what they are.

zero element. That element is the number 12. If we look back at Table 4, in section 3.1, we can see that whenever a number is added to 12, (using clock-addition) the result is that the number is returned. For example, $4 \oplus 12 = 4$ and $12 \oplus 7 = 7$, and in particular $12 \oplus 12 = 12$. This is exactly how the number zero behaves. Hence, we do not need to make any adjustments to this set or recheck it against any previous axioms (provided a thorough check was made previously). If we care to we could replace the symbol 12 with the symbol 0.

However, if we choose to use 12, then we need to be careful because the axioms may ascribe other properties to 0 that the number 12, as such, may not conform to. For example, $12/12 = 1$, while $0/0 \neq 1$. However, we will attend to that matter when we discuss multiplicative inverses.

All four of our systems, $[D,+]_A$, $[E_0,+]$, $[N_0,+]$ and $[C,\oplus]$, therefore, conform to the axioms so far presented.

~

3.5 Additive Inverses and Subtraction

Negative numbers* were not fully accepted by Western mathematicians as numbers until as late as the 17th century. (In the East, it was considerably earlier.) Prior to this time many mathematicians thought that negative numbers were counter intuitive, dark and obscure objects. Others believed that they just plain didn't exist and were nothing more than a false artifact of calculations. They were considered the absurd results of subtraction whenever the *subtrahend*, the number being subtracted, was greater than the *minuend*, the number from which a number was being subtracted, as in $52-67$ (52 is the minuend and 67 is the subtrahend). Thus, 2 pebbles take away 3 pebbles was considered nonsensical. Negative numbers do, indeed, seem to be more difficult to imagine than the natural numbers, but negative numbers like the natural numbers are also just ideas.

Similarly, as we have learned in high school (or perhaps not), negative numbers behave exactly like positive numbers, only negatively. Whenever we add natural numbers, which are positive (zero is considered neither positive nor negative), we expect to have a greater number of them. For example, $4+3=7$, and 7 is greater, positively (relatively) speaking, than either of the numbers we started with. We know this about addition in general. It is how addition works for all things that have the same characteristics. Thus, three apples added to four apples gives us more apples, in this case seven apples. It is the same with negative numbers. Adding negative numbers gives us more negative numbers. Negative numbers have the characteristic of being, well,

---

* Although it is convenient to call the numbers under discussion *negatives*, it is technically incorrect in many cases and can lead to confusion. This will become more apparent when we discuss the additive inverses of wrap-around arithmetic.

negative.* Hence, $(-4)+(-3)=-7$, and (-7) is greater, negatively (relatively) speaking, than either of the numbers with which we started.†

Normally, we cannot add things that have different characteristics without some definition: three apples and four frogs are neither seven apples nor seven frogs. The next axiom will define for us a way to add positive and negative numbers, but before we can do that we need some assurance that negative numbers exist. What good is being able to add positive and negative numbers, if we do not have any negative numbers—enter Axiom A5, which not only calls negative numbers, or more correctly additive inverses, into existence but also tells us how to add positive and negative numbers.

❖ **AXIOM A5: THE EXISTENCE OF ADDITIVE INVERSES**

For every number, a, there is a number called the additive inverse of a, written as $-a$, such that $a+(-a)=0$

This axiom, like Axiom A1, is recursive: if a is a number, then so is $-a$, but if $-a$ is a number, then so is $-(-a)$, but if $-(-a)$ is a ... well, you get the picture.

As it so happens, we already have an authentic number from the previous axiom, and that number is 0; consequently, we also have the authentic number, -0. Our discovery set would then consist of all pairwise additions of these two numbers (by A1). However, because of Axioms A2 through A4, many of the elements of this set are equal to each other; hence, it would reduce to the set $\{0,-0,-0+(-0),-0+(-0)+(-0) \cdots \}$, which is just the number 0 and a bunch of -0s added to each other (adding 0 to any of these numbers would, of course, have no effect on them).

There is no need to check if the elements of this set satisfy the conditions of all the previous axioms or for that matter those to come, it does because all of the elements *are* (authentic) numbers.

However, this axiom states that $0+(-0)=0$ and Axiom A4 states that $0+(-0)=-0$, then by the transitive property of equality (E3), it must be the case that $0=-0$. Hence, all the -0s and sums of -0s collapse into zero repeats and we are left with just the set we started with: $D=\{0\}$. (It should also be clear by substitution of equals (E6) that $-(-0)=-0=0$.)

In case you haven't notice, we have just proved the following theorem: *zero is its own additive inverse.* (Note: It *cannot* be proved within the confines of the axioms of algebra that any other number is equal to its additive inverse; 0 is

---

* Technically, all positive numbers should have the sign "+" preceding them in the same way the negative numbers have the sign " − " preceding them. Thus, to be technically correct, we should write $(+4)+(+3)=(+7)$ when dealing with positive numbers. However, by convention, the absence of a sign for numeral-numbers means that the number is to be considered positive. With that said, it is easy to conceptualize both positive and negative numbers, symbols are symbols.

† "greater" is being used here in a relative way, later we will look at it in an absolute way.

the only number with this property. [Wrap-around arithmetic is different because of the definitions involved in this type of arithmetic.])

◆ **THEOREM 3.5.1**

$0 = -0$

There is another property of numbers that we can prove with the axioms we currently have on hand, which is due to the recursive property of this axiom, and that is that *the additive inverse of the additive inverse of a number is the number.*

◆ **THEOREM 3.5.2**

$-(-a) = a$

Proof: $-a$ is a number by A5 (that is, if a is a number, which we assume it is), and therefore it too has an additive inverse (by this axiom), which is, accordingly, $-(-a)$* and it is such that $-a+(-(-a))=0$. Now, the addition property of equality, E4 (Chapter 2, Section 2.1), allows us to add the number a to both sides of this equality; hence, we may write $a+[-a+(-(-a))]=a+0$. The associative property, A3, allows us to shift the parentheses (the square brackets here) on the left hand side of this equality giving us $[a+(-a)]+(-(-a))=a+0$. However, $-a$ is the additive inverse of a according to A5; hence the addition of these two numbers is equal to 0. Therefore, we have $0+(-(-a))=a+0$ (the superfluous parentheses have been removed). However, 0 added to any number is the number according to A4; consequently, we have the result $(-(-a))=a$, and, after removing the superfluous parentheses, we have $-(-a)=a$, which is what we wanted to prove. Thus, a and $-a$ are additive inverses of each other.

There is still another useful theorem we can prove immediately, it states that *the additive inverse of a sum is equal to the sum of the additive inverses.*†

◆ **THEOREM 3.5.3**

$-(a+b) = -a+(-b)$

Proof: If a and b are numbers, then so is $a+b$ (A1), which, therefore, has an additive inverse $-(a+b)$ by A5. Accordingly, we may write the axiomatic equation $(a+b)+(-(a+b))=0$. Because a and b are numbers, they each have additive inverses as well (A5), and because the sum of these additive inverses is a number, we may add their sum to both sides of $(a+b)+(-(a+b))=0$ and obtain $[-b+(-a)]+[(a+b)+(-(a+b))]=[-b+(-a)]+0$ by the addition property of equality, E4. Shifting parentheses (that is, using A3 twice), we may continue with $[-b+((-a+a)+b)]+(-(a+b))=[-b+(-a)]+0$. Now, because a and $-a$ are additive inverses, their sum is 0 (A5), and because 0 added to any number is the number (A4), we have $[-b+b]+(-(a+b))=-b+(-a)$. Once again, because

---

*That is how we make additive inverses: we put a dash, "−", before the number.

† This pattern, where the words are reversed, is found everywhere in mathematics.

$-b$ and b are additive inverses, their sum is 0; by A5, and because 0 added to any number is the number (A4), we have $-(a+b)= -b+(-a)$. The right hand side of this equation is equal to $-a+(-b)$ by A2. Finally, by making use of the substitution property of equality (E6), we have, with a long awaited *voilà*, $-(a+b)=-a+(-b)$, which is what we wanted to prove.

The following corollary follows directly from theorems 3.5.3 and 3.5.2. (Extending Theorem 3.5.3 will be useful when we get to the distributive property.)

◇ COROLLARY 3.5.3

$-(a+b+c)=-a+(-b)+(-c)$, *and* $-(-a+(-b)+(-c))=a+b+c$

(We will assume that both parts of this corollary are true for four or more numbers as well, although an induction axiom, which we briefly discussed in Chapter 1 under Logic, would be required to prove it to be true for any number of numbers.)

Proof of the first part: $-(a+b+c)=-((a+b)+c)$ by axiom A3 (Theorem 3.5.3 only involves pairs of numbers, so we first need to introduce parentheses to create pairs of numbers.), then $-((a+b)+c)= -(a+b)+(-c)= -a+(-b)+(-c)$ (by using two counts of Theorem 3.5.3). (Proof of the second part is up to you. ☆7)

Using Axiom A5 from a practical point of view is rather straight forward. For example, to find the value of $-17+21$, we would re-write the expression as $-17+(17+4)=(-17+17)+4=0+4=4$. Although, I am sure that you have learned the addition of these two classes of numbers differently.*

Subtraction has been around for nearly as long as addition has. With the introduction of additive inverses, we can define subtraction in a more precise, mathematical way. Pictorially, it is the process of removing objects from a group of objects; we could show it in the following manner:

5–2=3

∘ DEF. 3.5.1: SUBTRACTION

For numbers a *and* b, *we define* $a-b=a+(-b)$

Thus, *subtraction of a number is the addition of the additive inverse of the number.*

Subtraction is used more often as a convenient shorthand rather than an as actual mathematical operation—it is easier to write $7-2-a-x$ than it is to write $7+(-2)+(-a)+(-x)$. Also, this definition *forces* the equality between two different configurations of symbols; without it, the sequence of symbols $a-b$ could not be realized—could not be deduced.

Having defined subtraction, we need to investigate how to properly use it. Firstly, subtraction is not commutative. We can prove this with a counter example: $7-2\neq 2-7$. That is to say, $7-2=7+(-2)=5+(2)+(-2)=5+0=5$, while

---

* Pedagogically speaking, the method used here better prepares a child for algebra.

$2-7=2+(-7)=2+(-2)+(-5)=0+(-5)=-5$. Hence, $a-b\neq b-a$, which is also the chief reason why we must read algebraic expressions and statements from left to right.

However, if we always used parentheses to communicate our ideas irrespective of any definitions or conventions, then mathematics could be performed with equal results from either the left or the right, there would be no confusion, but there would also be many distracting parentheses.

Secondly, subtraction is not associative. That is to say, $(a-b)-c\neq a-(b-c)$. As an example, $(10-6)-2=2$, while $10-(6-2)=6$. Reading from left to right (and ahead to the next operation)*, whenever we encounter parentheses, we perform the operations inside the parentheses first. If there are no parentheses, we perform the subtractions strictly from left to right. We may rearrange the subtractions if we like, but always with an eye to the definition of subtraction. Consequently, because $9-4-2=9+(-4)+(-2)$ and because addition *is* commutative, we may write $9+(-4)+(-2)=(-2)+9+(-4)$, from which it follows that $9-4-2=-2+9-4$.†

However, we currently have no authentic additive inverses to work with; -0, which we have proved is equal to 0, is its own additive inverse and is of little theoretical help. Therefore, we will postpone a deeper discussion about the addition of additive inverses (and about how positive and negative numbers are ordered) until such time that the axioms generate other additive inverses, not to mention any other numbers.

As to the matter of our four candidate systems, $[D,+]_A$, $[E_0,+]$, $[N_0,+]$ and $[C,\oplus]$, we have already shown that the system $[D,+]_A$ is, and will always be, in the running. As for $[E_0,+]$, we observe immediately that none of the numbers in the set E_0 are additive inverses, and we need an additive inverse for *every* number in the set. The only thing we can do is insert them and create a new set of even numbers from the old, which we will label as E_{0-} (Notice that a dash, " – ", has been added to the subscript to denote the addition of the negative numbers to this set.), to wit:

$$E_{0-}=\{\ \cdots\ -6,\ -4,\ -2,\ 0,\ 2,\ 4,\ 6\ \cdots\ \}.$$

I leave it to you to convince yourself that this set under ordinary addition is closed and conforms to all the axioms we have so far covered.

$[N_0,+]$ is in the same situation that the even numbers were in. Hence, we will supply it with an additive inverse for every number already in the set. There is no need to include the additive inverses of the additive inverses because by Theorem 3.2.2 they are just repeats of numbers already in the set. Our new system will be denoted by $[N_{0-},+]$ (again, note the dash).

---

* If the next operation is multiplication, then we must perform that first.

† Think of the dash (negative symbol) as being glued to the number following it: where the number goes, so goes the dash.

The case for $[C,\oplus]$ is different. Every number in the set C *does* have an additive inverse: $1\oplus11=12$, $2\oplus10=12$, $3\oplus9=12$ ... $11\oplus1=12$ and $12\oplus12=12$ (recall that 12 has taken the place of 0). Although these additive inverses do not have the typical dashes (negative sign) preceding them that we are accustomed to seeing with additive inverses that is irrelevant. It is enough that they behave like additive inverses and that we have one for every number in the set. (It is behavior that defines.)

However, not having negative clock-numbers presents a problem: how do we do wrap-around subtraction? For example, what is $10\ominus3$? (For reasons of consistency, we will use the symbol "$\ominus$" to denote wrap-around subtraction.) Using the definition of subtraction (Def. 3.5.1), we would write $10\ominus3$ as $10\ominus3=10\oplus(-3)$. Now, all we must do is figure out what number on the clock (-3) represents*. Recalling that (-3) is the additive inverse of 3 by Axiom A5, and that 9 is the clock-number we must add to 3 to get 12 (our 0), it follows that the additive inverse of 3 is 9, which we had already figured; in other words, $-3=9$.† Consequently, $10\ominus3=10\oplus(-3)=10\oplus9=7$. If we care to, we could define the inverses of every number on the clock in this way: $-1=11$, $-2=10$, $-3=9$, $-4=8$, $-5=7$, $-6=6$, etc. Wrap-around subtraction is now a "simple" matter. For example, $5\ominus11==5\oplus(-11)=5\oplus(-5\oplus(-6))=(5\oplus(-5))\oplus(-6)=12\oplus(-6)=12\oplus6=6$. Alternatively, because the additive inverse of 11 is 1, (i.e., $-11=1$), we could have written $5\ominus11=5\oplus(-11)=5\oplus1=6$.

Intuitively these answers make sense because going counter clockwise (the negative direction on the clock) 2 units would put us at 10 when going clockwise (the positive direction); i.e., $-2=10$. In words, this statement says that the additive inverse of 2 is 10. Likewise, going counter clockwise 3 units puts us at 9 when going clockwise, i.e., $-3=9$, etc. Of course, we didn't have to bother with any of this, we could have found out what $5\ominus11$ is by starting at 5 and going counter-clockwise 11 units, bingo, 4 o'clock! However, if we had done that we would have missed out on all that glorious algebraic logic.

I leave it to you to find $8\ominus5$, $9\ominus3$ and $4\ominus11$ for the system $[C,\oplus]$, or better still, create the wrap-around subtraction table for the system; use logic to do your calculations and check it with your intuition.

Except for $[C,\oplus]$, our candidate systems for the game have changed, as have their nomenclature owing to the introduction of the additive inverse numbers, they are now $[D,+]_A$, $[E_{0-},+]$, $[N_{0-},+]$, and $[C,\oplus]$.

~

---

* If you're wondering why we wrote "(-3)" instead of "$(\ominus3)$", it is because that is what the axiom tells us to write; putting a dash before the number is how we indicate additive inverses.

† We are using the language of mathematics here, we are not saying that a negative three equals nine, what we are saying is that the additive inverse of three is nine, hence $-3=9$. You have encountered this language before when we were talking, logically, about the additive inverse of zero, i.e., $-0=0$.

Chapter 3 Exercises

1. The set {0, 3, 5} is clearly not closed. However, we were able to create a closed set using the three numbers in the set by repeatedly adding the numbers in pairs, both among themselves and among each other (see Section 3.1). Explain why when this process is continued indefinitely, the set created is equal to the following:

   {0, 3, 5, 6, 8, 9, 10, 11, 12, 13 ... }.

2. Fill in the table below to create a closed set for the elements α, β, γ, δ, such that each element is used exactly four times. (The only requirement here is that the set is closed, and that each element is unique). Check your entries.

+	α	β	γ	δ
α				
β				
γ				
δ				

   a. Does the addition table that you created abide by any of the axioms of addition? (Remember: it is not important what addition means—it doesn't mean anything—it is defined by what it does and what you call it. However, you may wish to replace each of the characters with a number between 0 and 3 and use wrap-around addition to create your table.)

   b. The number of ways in which the table can be filled given these restrictions involves the mathematics of combinatorics and probability. Although, I will not go into the details of finding the number of ways in which this can be done, you may wish to investigate a solution on your own—it is what mathematicians do. As a primer, you may imagine all of the characters (there are four of each for a total of sixteen) placed in a hat and then try to determine the number of different sequences that can be drawn from the hat.

3. Is the set $\{c\}$, where $c+c=c$ and $c \cdot c = c$, closed? If c is a number, then what number do you think c equals?

4. Using just the commutative and associative axioms of addition in pairs, determine the number of ways in which the expression $((\alpha+\beta)+\gamma)+\delta$ can be written, such that they are all equal.

5. Prove that $((\alpha+\beta)+\gamma)+\delta=(\gamma+\beta)+(\delta+\alpha)$ by writing out the sequence of derivation.

6. Which is colder, $-20°\text{F}$ or $20°\text{F}$? Which is warmer? Which is less cold? Which is less warm? If you had to choose between either the adjective *warm* or *cold* to describe the temperatures, which would you choose if you lived

in Hawaii? Which would you choose if you lived in Antarctica (the average winter temperature in Antarctica is about $-50^{\circ}\mathrm{F}$)? Why? How does this relate to positive and negative numbers?

7. It is only a matter of convention that mathematicians do not put a + (positive) symbol before each positive number; although it requires remembering that fact, it also saves on a lot of writing. Suppose "negative" numbers had been chosen to be greater than "positive" numbers, would that have made a difference in the mathematics? Explain why or why not?
8. Without using any other axioms, theorems or definitions than those already presented, simplify (reduce to its smallest length) the expression. (Pair up parentheses, if necessary, to avoid mistakes.)
   a. $x+(-2)+(a)+(-(-4))+(-b)$
   b. $13+(-(-(-7)))+(-(-y))+(-b)+x$
   c. $-z+45+(-(-x))+(-a)$
   d. $-(-(-6+x+(-7))+(-y))$
9. Although all letters of the alphabet are used as logical variables or place holders for numbers (among other mathematical objects), mathematicians reserve the last few letters of the alphabet for those logical variables that are defined to be either *independent* of everything in an equation or *dependent* on everything in an equation, while the first few letters of the alphabet are reserved for logical variables that are considered *constant* or *fixed*, albeit unknown, numerals. In the following equations, show how the dependent variable depends on everything else in an equation by solving for x, y, or z, use only the axioms, theorems and definitions already presented.
   a. $-(a+7)-y=-14+b-a$ b. $c-(-0-(-(-17-d)))+c=x-(-c)$
   c. $a+b+(-z)-(a+b)+b=-(22+a)$
10. In the equation $y=3x+9$ freely choose four different numbers for the independent (logical) variable x and determine the number that the dependent (logical) variable y equals. Fill in the table with your results.

x				
y				

11. There are names for systems that conform to certain axioms. A system that (1) conforms to a closure axiom (A1 or M1 [see multiplication axioms]), (2) contains an identity element (A4 or M4), (3) conforms to the associative property (A3 or M3), (4) contains an inverse for every element in the set (A5 or M5), and (5) has one operator ($+$ or $\cdot$) is called a *group*. If the group also conforms to the commutative property (A2 or M2), it is called an Abelian or commutative group. Whatever other axioms the system does or does not conform to is immaterial. Given the systems $[D,+]_A$, $[E_0,+]$, $[N_0,+]$,

$[N,♣]$, and $[C,\oplus]$ as they stand up to this point, determine which are and which are not groups, and which are Abelian groups.

12. Is the infinite, closed set that we generated from the set {0, 3, 5} (see problem #1), a group? If it is not a group, what must be done to make it a group? Is the set generated from the elements of the set {0, 5, −5} a group?
13. Using wrap-around addition, where 12 o'clock is replaced by zero, generate a set under this type of addition from the given sets and determine if the generated set is a group under addition, if the generated set is a subset of C (the set of elements on a clock), and which, if any, of the generated sets are equal.
    a. {2} b. {3} c. {5} d. {6}
    e. {7} f. {9} g. {11}
14. Why do some of the generated sets of problem 13 (above) equal C and others do not? Suppose we had a fifteen-hour clock instead of a twelve-hour clock, which of the elements of that set would generate all fifteen numbers? Can you make any conjectures about the type of generated sets and the numbers on a clock, irrespective of how many hours are on the clock?

CHAPTER 4
THE AXIOMS OF MULTIPLICATION AND DISTRIBUTION

We now turn our attention to the operation known as multiplication, which will be represented by a dot placed halfway up between the two numbers being multiplied whenever there is a need to emphasize the operation of multiplication (which, as it turns out, will be often), otherwise the two numbers will be placed directly adjacent to one another. Thus, we could write any of 2×3, $2\cdot3$, $(2)(3)$, $a\cdot b$, $(a)(b)$, and ab to signify multiplication.

I should warn you, though, that I will be using the dot symbol over the traditional times symbol for multiplication quite often, even between numerals. For some of you this may be annoying, my apologies. The reason for this preference is twofold, (1) the symbol "$\times$" has an ingrained meaning to most of you, the effect of which only makes it more difficult for you to see multiplication from a purely logical, abstract perspective; (2) the symbol "$\times$" is rarely used in higher mathematics. All that aside, the appearance of the symbol that falls between two numbers is irrelevant, what matters is how it is defined and whether it is labeled multiplication or addition.*

If you take a moment and compare the axioms of addition to the axioms of multiplication, you will notice that they are *exactly*† the same in terms of structure. If in all of the five axioms of addition we were to replace every occurrence of $+$ with $\cdot$, every occurrence of 0 with 1, every occurrence of the word *add* with the word *multiply*, and every occurrence of $-a$ with $1/a$, then we would have all of the axioms of multiplication. We could, literally, copy and paste every sentence from Chapter 3, make the appropriate replacements and say as much about multiplication as has already been said about addition.

However, I will not waste your valuable time by repeating myself, except when it may provide clarity between the similarities and differences of the two operations or when it will contribute to your reading experience. Therefore, I may, on occasion, refer you to the appropriate section in the chapter on the addition axioms (Chapter 3) whenever an analogous concept presented in this chapter was discussed in that chapter.

The process of multiplication that we use today is rather complex, requiring a considerable amount of prior knowledge. Besides knowing the multiplication tables, the place value of the digits, when and what to carry to the next place value, and how to add the carry for the next multiplication, one must

* There is only one symbol for the operation of addition. Analogously, however, the plus symbol can also be thought of as meaning either traditional addition or logical, abstract addition.

† Well, with one minor difference, we may not write $1/a$ whenever $a=0$.

also know the addition algorithm. Figure 15 snapshots the complex process of multiplying the number 284 by 79. Unfortunately, more than a few children never really master it until well into eighth or even ninth grade.

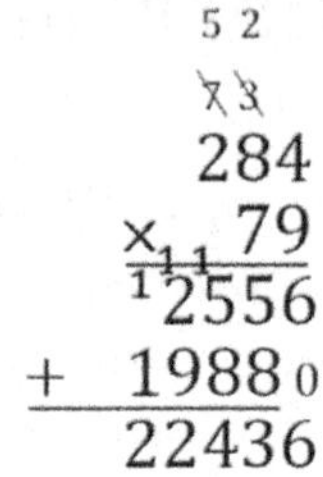

Figure 15: Modern Multiplication Algorithm

The Ancient Egyptians on the other hand, used a method that was easy to understand and certainly much quicker than adding 284 to itself 79 times, but not nearly as efficient as our modern method, partly because they did not use a place value system like we do. Instead they used an additive system, where each unique glyph (except their number 1) in any sequence of glyphs represented some multiple of ten. When ten counts of a symbol were reached, it was swapped out for the next higher glyph. Figure 16 shows the Ancient Egyptian number 284. (The arrangement of the glyphs is not particularly important because any arrangement would have been understood as the number 284.)

Figure 16: Ancient Egyptian portrayal of the number 284

However, without getting too deep into Ancient Egyptian mathematics, their method of multiplication was by way of a process called successive duplication or doubling as depicted (in modern decimal format) in Table 5.

A	B	C	D
Doubles of 1	Make 79	Doubles of 284	Seventy-nine 284's
1	1	284	284
2	2	568	568
4	4	1136	1136
8	8	2272	2272
16	-	4544	-
32	-	9088	-
64	64	18176	18176
~~128~~	*128 is beyond 79, so we stop at 64*		
Sum:	**79**	Sum:	**22436**

Table 5: Egyptian Style Multiplication

This process only works if we first have all the necessary doubles of 1 available for view or consideration. Then, working from the largest number, in this case 64, down to the smallest number, 1, we add those doubles that will "fit" into 79. The sum of 64 and 32 is too big, as is 64 and 16, but the sum of 64 and 8 fits comfortably within 79, as do all the numbers 64, 8, 4, 2, and 1, the sum of which is exactly 79 (Table 5, column B). When we gather up all the corresponding doubles of 284 (column D) and add them, we will have added seventy-nine 284s, which works out to be 22,436.

This doubling method was somewhat easier to do using hieroglyphics (Figure 17) than our modern decimal system because each glyph was simply repeated—doubled—however many times it needed to be after each doubling episode. However, it was also an extremely time-consuming affair because after each doubling episode the addition had to be completed and all of the replacements to the next numerically ten-fold-higher glyph had to be determined. Again, the abacus was an indispensable time-saving tool.

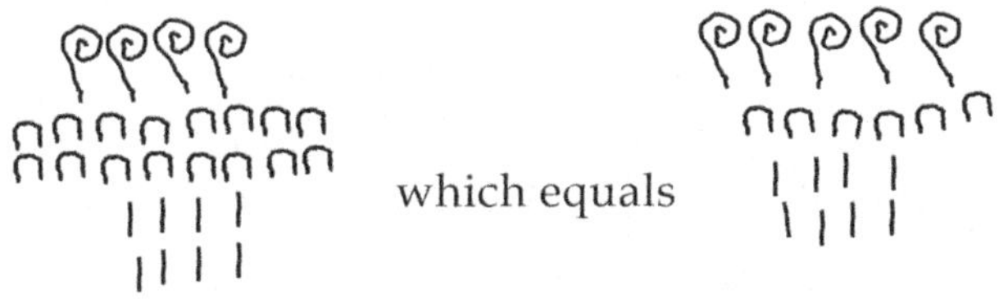

Figure 17: The double of the number 284 from Figure 16.

Regardless of their culture, civilized people seem to have recognized a connection between addition and multiplication, if only as a quicker way to do addition. We could say that multiplication is a type of shorthand for addition. This connection between the two operations culminates in the distributive axiom, which we will encounter at the end of this chapter.

However, in modern symbolic mathematics + and • * behave logically exactly alike and have no apparent connection whatsoever. Indeed, they would be completely independent of each other if it were not for the distributive axiom.

4.1 The Closure Property of Multiplication

As with addition, any eligible set of numbers to use with algebra must be closed under multiplication. With that in mind, I present the next axiom.

❖ Axiom M1: Closure under Multiplication

If a and b are (any) numbers, then ab is a number

Two numbers go into the machine and, according to Axiom M1, only one number comes out (see Figure 18 on the reverse) . . .

* or the more familiar symbol for multiplication " × " if you prefer.

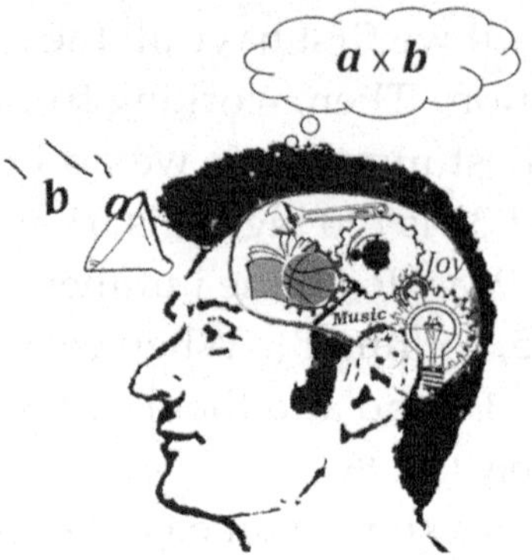

Figure 18: Mark (the multiplier) multiplies b with a

This axiom, like A1 and A5 before it, is recursive; thus, $(c\cdot(c\cdot(c\cdot(c\cdot c))))$ is a number if c is a number. The axiom only requires that we multiply numbers in pairs. Consequently, if a, b and c are any three numbers, then the following numbers can be constructed by the machine (and infinitely many others, of course):

$$ab,\ ba,\ ac,\ ca,\ bc,\ cb,$$
$$c(ab),\ c(ba),\ b(ac),\ b(ca),\ a(bc),\ a(cb),$$
$$(ab)c,\ (ba)c,\ (ac)b,\ (ca)b,\ (bc)a,\ (cb)a.$$ ▪4.1.1

However, augmenting these numbers and the numbers that were constructed from the closure axiom of addition (the numbers in 3.1.2) are the numbers the machine can construct from all pairwise combinations of a, b and c using *both* the addition and the multiplication closure axioms in succession, as shown below in 4.1.2.

	Col-1*	Col-2*	Col-3*	Col-4*
(Line 1a)	$a+(b\cdot c)=$	$(b\cdot c)+a,$	$a+(c\cdot b)=$	$(c\cdot b)+a,$
(Line 2a)	$b+(a\cdot c)=$	$(a\cdot c)+b,$	$b+(c\cdot a)=$	$(c\cdot a)+b,$
(Line 3a)	$c+(a\cdot b)=$	$(a\cdot b)+c,$	$c+(b\cdot a)=$	$(b\cdot a)+c,$
(Line 1b)	$a\cdot(b+c)=$	$a\cdot(c+b),$	$(b+c)\cdot a=$	$(c+b)\cdot a,$
(Line 2b)	$b\cdot(a+c)=$	$b\cdot(c+a),$	$(a+c)\cdot b=$	$(c+a)\cdot b,$
(Line 3b)	$c\cdot(b+a)=$	$c\cdot(a+b),$	$(b+a)\cdot c=$	$(a+b)\cdot c.$ ▪4.1.2

Because the closure axioms do not grant equality, we cannot make the claim that any of the numbers in either 4.1.1 or 4.12 are equal, except that for *each line* of numbers in 4.1.2 above, the numbers in columns 1 and 2 are equal by Axiom-A2, and the same is true for the numbers in each line in columns 3 and 4. (We have not yet considered the commutative or associative axioms of multiplication, thus we can make no further claims regarding equality.)

As you can see, the set of numbers that are constructible using either addition or multiplication or both, even from this limited selection of numbers, is getting rather large and a good bit complicated. If you recall from section 3.1

(The Closure Property of Addition), there were six ways to perform the addition of any two numbers from a collection of three (distinct) numbers, and twelve ways to perform the addition of all three numbers in a pairwise fashion. That was just for addition. With the inclusion of multiplication, we must double that count, and when we mix the operations of addition and multiplication together we will need to add another twenty-four (as displayed in 4.1.2) because when we mix the operations among three numbers, we may do any of

$$a+(b\cdot c),\ (a+b)\cdot c,\ (a\cdot b)+c, \text{ or } a\cdot(b+c)$$

for *each* arrangement of numbers. In other words, there are four ways for each of the twelve arrangements. Hence, the number of ways in which all three numbers, a, b and c, can be either added or multiplied or both, plus the number of ways any two of the three numbers can be added or multiplied, is sixty. Recognizing the many ways that numbers can be created will not only help you to perform algebra, but it will also help you to understand how the integers and the rational numbers originate from the axioms when we raise that discussion a little further on. Furthermore, higher mathematics abounds with these ideas of creating sets from the closure properties.

Before moving on to the next axiom, we must check our systems, $[D,+]_A$, $[E_{0-},+]$, $[N_{0-},+]$ and $[C,\oplus]$, for conformity to this new axiom. However, because we have introduced a new operation—multiplication, our current notation is no longer appropriate: it lacks acknowledgement of this important information. Accordingly, we will change it to include the multiplication operator: either $\bullet$ or $\times$. Our candidate systems to check will then be, $[D,+,\bullet]_A$, $[E_{0-},+,\times]$, $[N_{0-},+,\times]$ and $[C,\oplus,\otimes]$. (I am assuming for the moment that wrap-around multiplication will be different than ordinary multiplication.)

As usual however, there is no need to check whether $[D,+,\bullet]_A$ conforms to this axiom because, as mentioned, zero is a number given to us by algebra, and therefore any numbers we create from zero using the operators $+$ and $\cdot$ will also conform. However, for the moment, we are logically unable to determine what $0\cdot 0$ equals and therefore what it's additive inverse, $-(0\cdot 0)$, equals or what any combination of additions and multiplications of these numbers equal. (Except that by A5, we know that $0\cdot 0+(-(0\cdot 0))=0$, etc.)

Until we can determine what these numbers equal, our new set consists of 0 and a bunch of additions of products of 0s and of -0s:

$$D=\{0,\ 0\cdot 0,\ -(0\cdot 0),\ (0\cdot 0)\cdot 0,\ 0\cdot(0\cdot 0)),\ 0\cdot(-(0\cdot 0)),\ -(0\cdot 0)\cdot 0,\ (0\cdot 0)+(0\cdot 0)\ \cdots\ \}.$$

However, to simplify this ungainly set and future sets, we will make the following semi-formal assumption, which, unfortunately, we will not be able to prove until we get to the distributive axiom. It reads, *the product of zero and any number is zero.**

* If you prefer the long version, then *the product of the additive identity number and any number is*

◆ *Assumption #1: For every number* a, $a\cdot 0=0$ and $0\cdot a=0$ ◆

As a consequence of this assumption, because 0 is a number, we have $0\cdot 0=0$; via substitution of equals (E6) and Theorem 3.5.1, we also have $-(0\cdot 0)=-0=0$. Consequently, all of the elements in set D collapse to 0, and we are left with just the set $D=\{0\}$ all over again.

As for the candidates $[E_{0-},+,\times]$ and $[N_{0-},+,\times]$, I leave it to you to convince yourself that any pairwise multiplications of any of the numbers from within either one of the sets, E_{0-} or N_{0-}, will, itself, be a number from within that set. Therefore, they too remain candidates.

With regards to the system $[C,\oplus,\otimes]$, we first need to define what we mean by multiplication. We will define it in essentially the same "wrap-around" way that we did for addition, only for multiplication. Thus, for example, $2\otimes 8=4$ ($2\times 8=16$, which is 4 beyond 12), $9\otimes 3=3$ and $12\otimes 12=12$. We can create a multiplication table for wrap-around multiplication just as we did for wrap-around addition.

⊗	1	2	3	4	5	6	7	8	9	10	11	12
1	1	2	3	4	5	6	7	8	9	10	11	12
2	2	4	6	8	10	12	2	4	6	8	10	12
3	3	6	9	12	3	6	9	12	3	6	9	12
4	4	8	12	4	8	12	4	8	12	4	8	12
5	5	10	3	8	1	6	11	4	9	2	7	12
6	6	12	6	12	6	12	6	12	6	12	6	12
7	7	2	9	4	11	6	1	8	3	10	5	12
8	8	4	12	8	4	12	8	4	12	8	4	12
9	9	6	3	12	9	6	3	12	9	6	3	12
10	10	8	6	4	2	12	10	8	6	4	2	12
11	11	10	9	8	7	6	5	4	3	2	1	12
12	12	12	12	12	12	12	12	12	12	12	12	12

Table 6: Wrap-around multiplication for the number 12

~

4.2 The Commutative Property of Multiplication

From arithmetic, we know that the order in which we multiply two numbers will not change the result. We can visualize the truth of this statement using a grid of bells or the area of a rectangle, given some unit of measure.

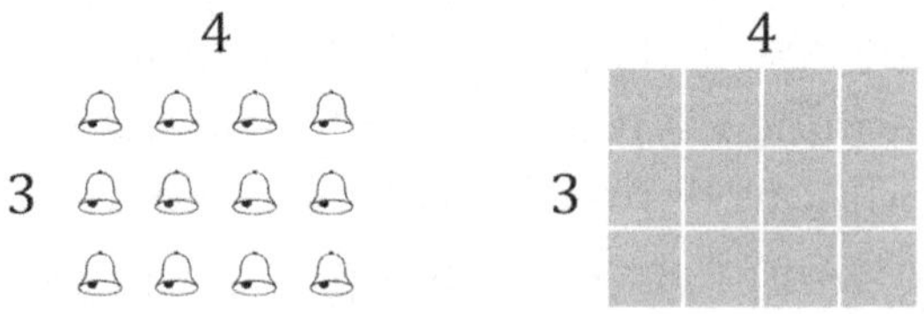

Figure 19: Commutative Property of Arithmetic

Whether we view the first image as three rows of four bells or four columns of three bells, the result is the same: twelve bells. Likewise, the same is true for the area of the second image involving 12 unit-squares.

---

the additive identity number.

By way of analogy between the structure of the axioms of multiplication and those of addition, we can conclude based on this axiom that $ab = ba$, $ac = ca$, and $bc = cb$, and that the numbers in each of the following groups (Figure 20) are equal:

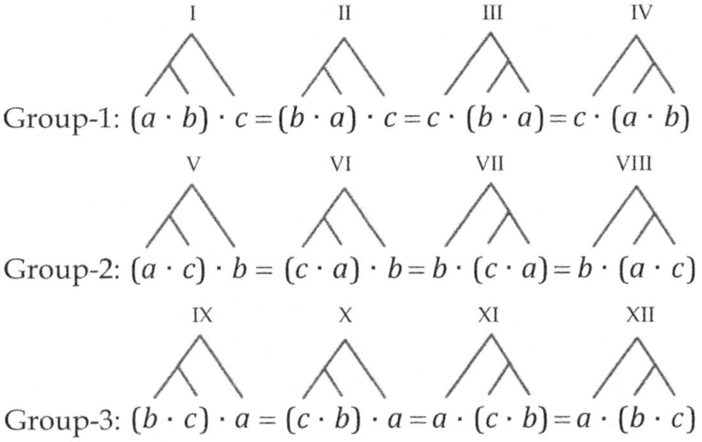

Group-1: $(a \cdot b) \cdot c = (b \cdot a) \cdot c = c \cdot (b \cdot a) = c \cdot (a \cdot b)$

Group-2: $(a \cdot c) \cdot b = (c \cdot a) \cdot b = b \cdot (c \cdot a) = b \cdot (a \cdot c)$

Group-3: $(b \cdot c) \cdot a = (c \cdot b) \cdot a = a \cdot (c \cdot b) = a \cdot (b \cdot c)$

Figure 20

As it was for the commutative property of addition, the commutative property of multiplication partitions the above twelve numbers into three distinct groups, only this time the numbers are composed of factors instead of terms. (Factors are to multiplication, what terms are to addition. Thus, $a+b+c$ consists of three terms, while $a \cdot b \cdot c$ consists of three factors.)

Also, all of the numbers in the same line of the collection of numbers in 4.1.2 are equal because each line consists of the same pairs of numbers, only this time there is one pair each for addition and multiplication.

However, the associative axioms of either addition or multiplication, which require at least three numbers to execute, will not cause any of the numbers from different lines of 4.1.2 to be equal because shifting parentheses is strictly confined to either addition or multiplication—not both. For that reason, it would be completely illegal and beyond the bounds of the axioms to write $(a+b) \cdot c = a+(b \cdot c)$. In other words, $(a+b) \cdot c \neq a+(b \cdot c)$.

However, there is something distressing about the interplay of the commutative axioms of addition and multiplication, which has left many students bald before their time. However, before I continue, I suppose I should first dispense with the commutative axiom of multiplication.

❖ Axiom M2: Commutativity of Multiplication

If a and b are numbers, then $ab = ba$

Here, then, is the cause of so many falling follicles: the commutative axiom of addition (A2) states that $a+b=b+a$ and the commutative axiom of multiplication (M2) states that $ab = ba$. However, given that the axioms of addition are virtually identical to the axioms of multiplication*, and that symbols, held together by logic, are still just symbols, what, then, are we to make of the simple, but ambiguous expression

$$a+b\cdot c\,?$$

The problem has to do with choice. Which (if not both or neither) of the two commutative axioms are we allowed to perform? The commutative axiom of addition states that

$$\underline{a+b}\cdot c=\underline{b+a}\cdot c, \qquad \blacksquare 4.2.1$$

whereas the commutative axiom of multiplication states that

$$a+\underline{b\cdot c}=a+\underline{c\cdot b}\,. \qquad \blacksquare 4.2.2$$

In arithmetic, we were taught, and we take it for granted that whenever we see an expression like $7+4\times 5$, we are expected to do the multiplication first and then the addition. Thus, $7+4\times 5=7+(4\times 5)=7+20=27$; we would never do the addition first—we would never calculate $7+4\times 5$ to be equal to $(7+4)\times 5=11\cdot 5=55$. However, because of the identical nature of the axiomatic structures of these two operations, there is no logical reason why it should have preferential status.

Furthermore, we were taught that we are allowed to write $a+b\cdot c=b\cdot c+a$, and not allowed to write $a\cdot b+c=b+c\cdot a$, where the only difference between the two expressions is that the pluses and the times symbols were swapped. However, if we are allowed to do the "move" that we did in the first expression for addition, then logically we should be allowed to do the same move in the second expression for multiplication because these two operations behave in (almost) every respect exactly alike, and the next four axioms, M3, M4, M5, and M6 (the distributive axiom), will not change that fact. (Although M5 does make the slight deviation from the symmetry between these two operations.) Yet, preference, in this case, is given to addition.

There is nothing, however, that has been said in any of our discussions so far that would grant preferential treatment to either the "dot" symbol, ·, or the cross symbol, "+", and it needs to be straightened out.

In all fairness, however, I need to point out that there isn't a problem here at all, and I apologize if I've been keeping you on the edge of your seat. (Did I say that out loud?) The reason that the "problem" arises at all is because the expression $a+b\cdot c$ is technically invalid. It is not deducible—not derivable—from the axioms of algebra. Recall that addition and multiplication are always carried out in pairs; consequently, the closure axioms will only allow us to produce either the number $(a+b)\cdot c$ or the number $a+(b\cdot c)$,† but they will not allow us to produce the number $a+b\cdot c$, which has no distinguishing pair-wise

---

* Except for the one minor blip where $a\neq 0$ whenever we write $1/a$.

† For the given sequence of numbers, $a\rightarrow b\rightarrow c$, and sequence of operators, $+\rightarrow\bullet$.

parentheses. Furthermore, which of the two deducible-numbers is produced depends on how the numbers, *in pairs*, and the operations of addition and multiplication go into the human calculating machine, and what she or he wants to communicate.

If we always used parentheses as the axioms call for, there would never be a problem trying to sort out whether one should do addition first or multiplication first—it would always be perfectly clear: do what's in parentheses first.

The reason we have the expression $a+b\cdot c$ is because mathematicians are lazy, and they would prefer not to write all those parentheses if they did not have to. So, mathematicians have made a choice, by convention they have decided that $a+b\cdot c$ shall be defined to be equal to $a+(b\cdot c)$. However, mathematics would work equally well, if they had chosen to define $a+b\cdot c$ as being equal to $(a+b)\cdot c$. It is, after all, just a definition, what matters is what one is trying to communicate, which is precisely what parentheses allow us to do.

From an algebraic point of view, this means that whenever we come across a "lackadaisical" expression like $a+b\cdot c$ or $a\cdot b+c$, then we are to assume that the writer meant that the two numbers separated by the "times" operator shall take precedence over the two numbers separated by the "plus" operator.

From an arithmetical point of view, it means that if we have an expression like $2+3\times5$ or $7\times8+12$, then we are required to perform the multiplication first to arrive at the correct calculation the writer had in mind. If the writer had anything else in mind, then she or he should have used parentheses. That brings us to our first definition pertaining to the order of operations.

∘ DEF. 4.2.1: ORDER OF OPERATIONS: MULTIPLICATION OVER ADDITION
The commutative property of multiplication shall take precedence over the commutative property of addition in the absence of parentheses; thus for numbers a, b, *and* c, *we define* $a+b\cdot c=a+(b\cdot c)$ *and* $a\cdot b+c=(a\cdot b)+c$.

Consequently, we may write $a+bc+d=a+(bc)+d=a+(cb)+d=a+cb+d$, and also $7\times2+3\times5=(7\times2)+(3\times5)=(2\times7)+(5\times3)=14+15=29$.

Although the definition is straight forward and should not present a problem, we may wish to extend it, in which case, we must be able to recognize whether any given mathematical expression is an expression of addition or an expression of multiplication, which itself requires being able to recognize the many ways that numbers can be depicted. We will discuss this matter in more detail in the next section because we will need the associative property of multiplication to develop that discussion. We will also investigate whether our candidate systems conform to this axiom in the next section as well.

~

4.3 The Associative Property of Multiplication

The associative property for multiplication states that regardless of how we group numbers in pairs for multiplying the results will be equal. We can represent it pictorially as in Figure 21.

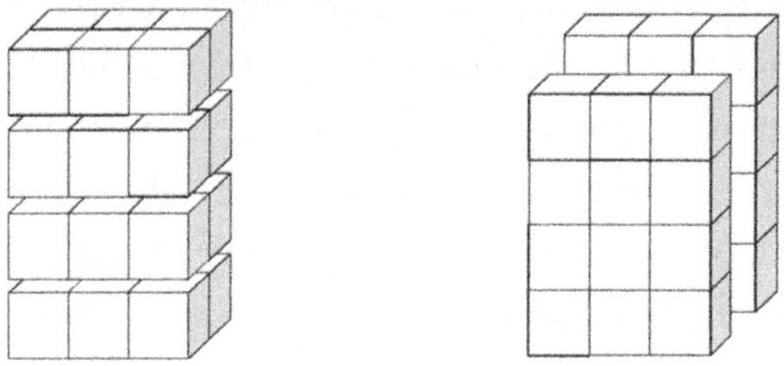

Figure 21: Pictorial representation of the associativity of multiplication

❖ Axiom M3: Associativity of Multiplication

If a, b and c are numbers, then $a(bc)=(ab)c$

With the associative and commutative axioms of both multiplication and addition we can consolidate our results from lines 3.1.2 and 4.1.1 and write that if a, b and c are any three distinct numbers, then the addition or multiplication of any two of them yields

$$a+b,\ a+c,\ b+c,\ ab,\ ac,\ bc. \qquad \blacksquare 4.3.1$$

Likewise, we may operate on all three numbers using either addition, multiplication or both to produce the eight distinct numbers

$$a+b+c,\ abc,\ a+bc,\ b+ac,\ c+ab,\ (a+b)c,\ (b+c)a,\ (a+c)b. \qquad \blacksquare 4.3.2$$

Any other arrangement or use of parentheses, will just be equal to one of these fourteen numbers. In other words, because of the commutative and associative axioms of these operations, we can reduce the sixty numbers that were constructed directly from the closure axioms to just fourteen distinct numbers.

That information, in and of itself, is not important. What is most important, however, is recognizing that there are a multitude of ways that we can express ourselves using the language of mathematics.

You will notice that there is no ambiguity in the order in which the operations for any of the numbers in 4.3.2 are to be performed, even though the 3rd, 4th, and 5th numbers in that line have no parentheses. Def. 4.2.1 has allowed us to remove those parentheses without loss of understanding.

However, as an example of an extension to Def. 4.2.1, to continue our earlier discussion from the previous section, suppose a mathematician wrote

$$a+(b+c)d \qquad \blacksquare 4.3.3$$

and we wanted to know what he meant by it? First, recall that such an expression is not derivable from the axioms of algebra (it is missing a pair of parentheses), hence it must mean one of either $a+[(b+c)d]$ or $[a+(b+c)]d$, which *are* derivable from the axioms, but which express completely different mathematical thoughts* and cannot be proved equal. However, the nested expression $(b+c)d$ in 4.3.3 is clearly understood as "multiply the number d by the sum of b and c", where the sum, $(b+c)$, is logically understood as one number (Axiom A1)†. In other words, the expression $(b+c)d$ is the product of exactly two numbers. Therefore, upon comparing the expression in 4.3.3 with Def. 4.2.1, the mathematician's meaning is clear, we should evaluate the expression $a+(b+c)d$ as $a+[(b+c)d]$ because multiplication takes precedence over addition in the absence of parentheses. Thus, $5+(3+2)\times 6 = 5+((3+2)\times 6) = 35$.

Suppose, on the other hand that our mathematician wrote $a+bcd$. Here again, the nested expression bcd can be thought of as the product of two numbers by virtue of the associative axiom of multiplication, then, after applying the order-of-operations definition, we have $a+bcd = a+(bc)d = a+((bc)d)$. As you can see, by convening and agreeing on the supremacy of multiplication over addition, mathematicians have removed many of the parentheses from mathematical expressions that, in an ironic twist, would otherwise clutter our view and obscure quick communication.

I have often found that one of the difficulties that students have with evaluating an expression is being able to distinguish between the major operation of an expression and the order of operations. For example, in the expression xy, multiplication is not only the major operation, it is the only operation—we are multiplying two numbers. However, in the expression $xy+z$, addition is the major operation—we are adding two numbers. In other words, this is an addition problem. Students often consider this latter expression to be a problem in multiplication because they have the phrase *multiplication comes before addition* ringing in their ears from years of drill.

Although it is true that we perform multiplication first and addition last, it is the *last* operation performed, irrespective of whether it is addition or multiplication, that distinguishes the type of problem any expression is. Thus, $(ax+b)(cy+d)+(z+n)(h+k)$ is an addition problem of two terms. The first term consists of two factors, the first of which consists of two terms, the first term of which consists of the two factors, a and x, etc., (ugh!). Thus,

$$\underbrace{\underbrace{\underbrace{(\underbrace{ax}_{\text{1st term}} + \underbrace{b}_{\text{2nd term}})}_{\text{1st factor}} \cdot \underbrace{(cy+d)}_{\text{2nd factor}}}_{\text{1st term}} + \underbrace{\underbrace{(z+n)}_{\text{1st factor}} \cdot \underbrace{(h+k)}_{\text{2nd factor}}}_{\text{2nd term}}}_{\text{Addition problem}}$$

Notice that we cannot perform the final addition until we figure out what the 1st term equals, and we cannot figure out what the 1st term equals until we

---

* It is easy to see how a student might, *wrongly*, shift the square brackets in $a+[(b+c)d]$ to the square brackets in $[a+(b+c)]d$.

† To repeat, every axiom is a logical statement about the noun "number".

figure out what the product of its two factors equals, and so on.

On the other hand, the expression $[(ax+b)+(cy+d)]\cdot[(z+n)+(h+k)]$ is a multiplication problem of two factors. The first factor consists of the two terms $(ax+b)$ and $(cy+d)$, and so on.

There are other properties about numbers that we may deduce using the associative axiom of multiplication. However, most of them have already been explored from the perspective of the associative axiom of addition. You may wish to re-familiarize yourself with some of those properties (Section 3.3), but with an eye towards the operation of multiplication instead of addition.

As for our candidate systems, we know that $[D,+,\bullet]$ conforms to both the commutative and associative axioms of multiplication because 0 is a number.

I leave it to you to convince yourself that the other three systems also conform to both axioms M2 and M3. (The wrap-around multiplication table for the number 12 can be found in Table 6 on page 82.)

~

4.4 The Multiplicative Identity Element, 1

Although it is often said that 1 is the loneliest number, nothing could be further from the truth. 1 is the number that primitive humans first romanticized about. 1 is the number from which every other number gets its bearings. 1 is the yin to 0's yang: existence and non-existence. Even if numbers were nothing more than a brilliantly devised construct of human imagination—a naive attempt to mask our greatest fear that the world is totally beyond logic and reason, the numbers 1 and 0 would nevertheless shine through, locked in an eternal struggle over supremacy.

If, as the ancient sages believe, everything from the vacuum of space, which is one thing, to everything in it, which is another, is One (i.e., one thing), then it must also be nothing—it must also be Zero. Things only exist in relation to something else. If in all the vastness of vacuous space there existed nothing but one solitary, pure-white marble, then it would spontaneously "vanish"—become boundless—before the mind's eye in the absence or removal of the space surrounding it. In other words, everything is nothing, and I am not just playing with words: if everything is One thing, then everything (i.e., that *One* thing) is Nothing.

My apologies, I digress—back to more "earthly" matters . . .

❖ **Axiom M4: The Existence of the Multiplicative Identity Number**

There exists a number called one, which is written as 1, *such that if* a *is a number, then* $a\cdot 1=a$ *

Firstly, it is interesting to note that because 0 is a number, it follows from

* I leave it to you to show that the number 1 is unique (*hint: see the proof for the uniqueness of* 0).

this axiom that $1 \cdot 0 = 0$. This may come as a surprise to many of you. In other words, from a logical point of view, one times zero is zero not (necessarily) because of any property that zero possesses, but because of the property that the number one possesses: 1 returns the identity of every number it multiplies and 0 is a number—by axiom. Here 1 is the active agent, while 0 is passive. To put it another way, we must shift (for the moment) our way of thinking: we have *one zero* and not *zero ones*. However, a little later, we will be able to prove that zero times any number is zero, at which point zero will become the active agent. Of course, at that point we will have also demonstrated what $0 \cdot 0$ equals, and we will have also justified our assumption regarding the product of zero with any number (Assumption #1).

Secondly, this axiom delivers up another authentic number from which, employing Axioms A1, M1 and A5 (and later M5), we can amass a whole slew of other authentic numbers. However, until we determine what these numbers look like (which we will do presently), our discovery set, *D*, is just the axiomatically given numbers 0 and 1, in other words $D = \{0, 1\}$.

Because 1 is a number, it follows from this axiom, and Axiom M1 and the substitution axiom of equality (E6), that $1 \cdot 1 = 1$, $1 \cdot (1 \cdot 1) = 1$, $(1 \cdot 1) \cdot (1 \cdot 1) = 1$, etc., and that all parenthetical pairwise multiplications of 1 with itself are equal to 1. In this regard, 1 in relation to multiplication behaves exactly as 0 does in relation to addition, and therefore the set $D = \{0,1\}$ is unchanged by any of these products.

However, when we consider just the number 1 in relation to addition, it is not at all clear what the number $1+1$ *logically* equals. Neither is it clear what $(1+1)+1$ equals nor what any number of 1s added together is equal to. However, because 1 *is* a number (again, no *if*s about it), it conforms to all of the axioms of algebra, including in particular A2 and A3. Hence, we may dispense with all of the parentheses in any parenthetical pairwise additions of 1 and simply ask, what do the following additions equal?

$$1+1,\ 1+1+1,\ 1+1+1+1,\ 1+1+1+1+1,\ \text{etc.}^{*} \qquad ▪4.4.1$$

As it turns out, these sums of ones, unlike the products of zeros, $0 \cdot 0$, $0 \cdot 0 \cdot 0$, etc. (recall that by Assumption #1, 0 times any number is 0), are not equal to some other number that we are currently aware of in our discovery set $D = \{0, 1\}$, they certainly don't look like anything in that set, but they are decidedly in the set by Axiom A1. However, no theorem or axiom that we have so far encountered, or will encounter for that matter, will ever prove that they are anything other than what the eye can see: a visually unique string of 1s and +s, each differing from the next only in the count of each. What all this means is that the set *D* must necessarily expand to accommodate these new numbers.

---

* The symbols 2, 3, 4, etc., have not yet been defined.

However, we can make these numbers more manageable by giving them correspondingly unique names or labels. We will label $1+1$ with the symbol 2 or, logically speaking, we will define $1+1=2$. We will label $1+1+1$ with the symbol 3 or rather we will define $1+1+1=3$. We can define all of these visually unique strings of symbols accordingly: $1+1+1+1=4$, $1+1+1+1+1=5$, and so on. Thusly, we have generated the positive integers from the axioms of algebra, the order of which, for the moment, is only assumed*.

∘ DEF. 4.4.1: THE SET OF POSITIVE INTEGERS

$Z_P = \{\ 1,\ 1+1=2,\ 1+1+1=3,\ 1+1+1+1=4,\ 1+1+1+1+1=5,\ \ldots\ \}$

Of course, these symbols, 1, 2, 3, etc., have a history, which corresponds nicely to the axiomatic construction of these numbers.

Moreover, because we are given 1, we also have its additive inverse by Axiom A5, namely -1. Thus, as it was with the number 1, all of the following numbers, which involve only the number -1 and the operation of addition, are also numbers:

$$-1,\ -1+(-1),\ -1+(-1)+(-1),\ -1+(-1)+(-1)+(-1),\ \text{etc.} \qquad ▪4.4.2$$

Because any number of 1s added with an equal number of –1s can be shown to be equal to zero (recall section 3.5), it follows that any additive mix of the numbers from line 4.4.1 and 4.4.2 will ultimately be a number from exactly one of these two groups of numbers depending upon whether there are more 1s or more –1s. For example, the number $-1+(-1)+(-1)+1+1+(-1)$, can be shown to be equal to $-1+(-1)$, which is in the group of numbers from 4.4.2.

We can simplify the writing of the numbers in 4.4.2 with definitions just as we did for the numbers in 4.4.1, and in doing so construct the negative integers (the order is assumed).

∘ DEF. 4.4.2: THE SET OF NEGATIVE INTEGERS

$Z_N = \{\ldots(-1)+(-1)+(-1)+(-1)=-4,\ (-1)+(-1)+(-1)=-3,\ (-1)+(-1)=-2,\ \ -1\}$

The union of the positive and negative integers and zero is the entire set of integers. The set of integers is most often represented with the letter Z, from the German word for numbers: *zahlen*. Accordingly, we have the following definition (again the order is only assumed):

∘ DEF. 4.4.3: THE SET OF INTEGERS

$Z = Z_P \cup Z_N \cup \{0\} = \{\ldots\ (-1)+(-1)(-1),\ (-1)+(-1),\ -1,\ 0,\ 1,\ 1+1,\ 1+1+1\ \ldots\} =$
$\{\ \ldots\ -3,\ -2,\ -1,\ 0,\ 1,\ 2,\ 3,\ \ldots\ \}$

It is easy enough to verify that each of the negative integers is the additive inverse of some positive integer, for example,

* Order involves the symbol "<", and we haven't, yet, said a word about it (besides the order axioms).

$$\underbrace{2+(-2)=(1+1)+[(-1)+(-1)]}_{\text{Def. of 2 and }-2}=\underbrace{1+[1+(-1)]+(-1)}_{\text{Assoc.}}=\underbrace{1+0+(-1)}_{\text{Add. Inv.}}=\underbrace{1+(-1)=0}_{\text{Zero \& Add. Inv.}}.$$

We can also explain why $2+2=4$: $2+2=(1+1)+(1+1)$ by definition and by the substitution property of equality. However, $(1+1)+(1+1)=1+1+1+1$ by A3, and $1+1+1+1=4$ by definition, then using two counts of the transitive property of equality, we have $2+2=4$.

Of course, on another level there is no need to explain why two plus two equals four, because the truth of that statement predates modern mathematicians' need for axiomatic structures by many millennia. However, if the modern axiomatic view is going to coincide with what people, both ancient and modern, have always known about numbers, then the axioms of algebra must be able to parallel and reproduce this ancient knowledge precisely.

As another example, we can calculate $2-5$ as follows: $2-5=2+(-5)$, because that is how we define subtraction (Def. 3.2.1). However, $2+(-5)=(1+1)+[(-1)+(-1)+(-1)+(-1)+(-1)]$, which we arrived at by using the definition of the numbers 2 and -5, and the substitution of equals property (E6). However, $(1+1)+[(-1)+(-1)+(-1)+(-1)+(-1)]=(-1)+(-1)+(-1)$ after having applied A3, A5 and A4 a few times, and $(-1)+(-1)+(-1)=-3$ by definition of the symbol -3. Finally, we may conclude that $2-5=-3$, by the transitive property of equality. In a most elegant fashion, we have deduced much of basic arithmetic from the axioms of algebra.

Although some mathematicians may scoff, especially the likes of Russell and Whitehead * (if they were alive today), at the following proof that states the sum of any two integers is an integer, we will, nevertheless, present it as a (very) credible argument for that statement.

◆ THEOREM: SUM OF INTEGERS

The sum of two integers is an integer

We will begin by supposing that the first integer of the sum of two integers consists of j terms, all of which are only the number 1 or only the number -1. Thus, the first integer is either positive or negative. We will label the first integer as A. The second integer will consist of k terms, all of which are also only the number 1 or the number -1. Thus, this integer is also either positive or negative. We will label the second integer with B. In the case where A and B are both positive integers, we have

$$A+B=(\underbrace{\overset{\text{1st}}{1}+\overset{\text{2nd}}{1}+\overset{\text{3rd}}{1}+\overset{\dots}{\dots}+\overset{j\text{th}}{1}}_{\text{1st integer of } j \text{ terms of 1}})+(\underbrace{\overset{\text{1st}}{1}+\overset{\text{2nd}}{1}+\overset{\text{3rd}}{1}+\overset{\dots}{\dots}+\overset{k\text{th}}{1}}_{\text{2nd integer of } k \text{ terms of 1}}).$$

The associative axiom of addition allows us to remove these parentheses, from which we may write (overleaf),

* PM required some one hundred pages to prove that $1+1=2$

$$\underbrace{\overset{1\text{st}}{1} + \overset{2\text{nd}}{1} + \overset{3\text{rd}}{1} + \overset{\dots}{\dots} + \overset{j\text{th}}{1}}_{j \text{ terms of } 1} + \underbrace{\overset{j+1\text{st}}{1} + \overset{j+2\text{nd}}{1} + \overset{j+3\text{rd}}{1} + \overset{\dots}{\dots} + \overset{j+k\text{th}}{1}}_{k \text{ terms of } 1}$$

However, this is a string of 1s and +s, which differs from other such strings only by the count of each. Therefore, even without knowing what the sum is equal to, it has the form of an integer, and therefore is an integer. I will leave it to you to consider the case where A and B are both negative (the case where one or both are zero is trivial).

In the case when one integer is positive, say A, and the other negative, B, we would have the following:

$$A+B=(\underbrace{1+1+1+\ \dots\ +1}_{\text{1st integer}})+(\underbrace{-1+(-1)+(-1)+\ \dots\ +(-1)}_{\text{2nd integer}}).$$

After which, using the associative and additive inverse axioms to group the numbers in such a way to create zeros, we obtain

$$A+B=1+1+\ \dots\ +[1+\underbrace{[1+\underbrace{[1+(-1)]}_{=0}+(-1)]}_{=0,\text{ etc.}}+(-1)]+\ \dots\ +(-1)+(-1).$$

As can be seen, the zeros will "vanish" from the inside out, and we will be left with either a string of positive ones or a string of negative ones, depending upon which of the two integers had the greater count. Regardless, in the end, it will have the form of an integer.

I am hoping that I am not losing my audience up to this point, but I should make some clarifications just to be sure. All of the numbers that logically arise from the axioms, are of course "authentic" numbers (as I have been using the term). However, there are different classifications of numbers. The integers, for example, are a certain subclass of numbers, and because the integers logically arose from the axioms, the sum of any two integers will, by axiom, be a *number*; however, the sum of any two integers may not necessarily be an *integer number* (although, we have just shown that it is).

The same is true for the product of two integers. The product of two integers is, by axiom, a number, but not necessarily an integer number. In each case, we would have to prove closure within the set of *integers* if we wish to determine the minimal subset of numbers that the axioms produce. When we introduce the concept of multiplicative inverses, the same sort of questions will be discussed over those numbers as well.

Furthermore, the point of all this, if some of you are still a bit foggy, is to determine where the axioms logically lead. By that directive and because some numbers have originated from the axioms, we are obliged to complete the set of numbers deducible from them. I hope that has clarified a few things for anyone who may have been having some difficulties with the discussion so far presented, but back to the discussion at hand.

Although it has been a rather straight forward process generating the integers using addition and the numbers 1, −1, and 0, it is a little more problematic determining what the product of any of the integers are, in particular we would like to know what $(-1)\cdot(-1)$ is equal to, and more generally what (-1) times any number is equal to. However, for the sake of simplicity and for want of the distributive axiom, we will make a second, semi-formal assumption: whenever we multiply any number by (-1), we obtain the additive inverse of the number, in math terminology we would write

◇ *Assumption #2:* $(-1)\cdot a = -a$ ◇

Under this assumption then, $(-1)\cdot(-1)=1$ because 1 is (-1)'s additive inverse (i.e., $1+(-1)=0$). Also, (-1) times any integer from either set Z_P or set Z_N would then change it to its additive inverse and Corollary 3.5.3a* would then place that number in the opposite set. For example,

$$\underset{3\in Z_p}{(-1)(3)} = \underset{\text{Assumption \#2}}{(-1)(1+1+1) = -(1+1+1)} = \underset{\text{Corollary 3.5.3a}}{(-1)+(-1)+(-1)} = \underset{-3\in Z_N}{-3}.$$

Also, under this assumption we can determine the sign of the product of any number of negative integers. For example,

$$-2(-3)(-4) = (-1)\cdot 2\cdot(-1)\cdot 3\cdot(-1)\cdot 4 = [(-1)(-1)]\cdot(-1)\cdot 2\cdot 3\cdot 4 =$$
$$1\cdot(-1)\cdot 2\cdot 3\cdot 4 = -1\cdot(2\cdot 3\cdot 4) = -(2\cdot 3\cdot 4). \qquad \text{■4.4.3}$$

Thus, the product of three negative numbers is the negative of the product of three positive numbers.

It should be clear that when a product consists entirely of an odd number of negative integers, the result is a negative integer, and when it consists of an even number of negative integers, the result is a positive integer.

However, what we cannot determine is what $2\cdot 3\cdot 4$ equals. Recall that the symbols 2, 3, and 4, are used only as definitions for the integers—an abbreviation of sorts for the actual integers $(1+1)$, $(1+1+1)$, and $(1+1+1+1)$. Therefore, the product of these three integers is actually represented by the number

$$(1+1)\cdot(1+1+1)\cdot(1+1+1+1).$$

Of course, this doesn't look anything at all like an integer because no integer has the multiplicative dot anywhere in its representation. An integer is 1, −1, 0, a sequence of 1s and pluses, or a sequence of −1s and pluses, and nothing else. Consequently, the product of any combination of numbers from the set of integers is semantically (and visually) different from the integers, if not logically different; in other words, they are not integers, at least from appearances, and appearances count. If they are integers, we must prove that we can make them look like integers, or at the very least explain why they are integers. However, this will require the distributive axiom of Section 4.6.

For those of you who are familiar with the distributive axiom, and I suspect that most of you are (for those of you who are a little shaky with it, may I

---

* i.e., $(-1)\cdot(a+b+c) = -(a+b+c) = -a+(-b)+(-c)$.

suggest you take a moment and review it now), I will present a short verification of the fact (which will be proved later) that the product of any pair of integers is indeed an integer:

$$\underbrace{\overset{2}{(1+1)}\cdot\overset{3}{(1+1+1)}}_{\text{Product of two integers}} = \underbrace{(1+1)\cdot 1+(1+1)\cdot 1+(1+1)\cdot 1=(1+1)+(1+1)+(1+1)}_{\text{Using the distributive property and the property of 1}} =$$

$$\underbrace{\overset{6}{1+1+1+1+1+1}}_{\text{Associativity of Addition}},$$

which happens to be the integer defined as 6.

If you have been able to follow the discussion up until now, then let me assure you that it has not been in vain. With all that has been said, it should be clear that we have constructed the set of integers with the axioms we have so far encountered using nothing more than logic and a smidgen of good common sense.

It follows that our discovery set, D, which we have been constructing from the axioms, consists entirely of the all the numbers called the integers; in other words, it is equal to the set Z. We have also shown that the set Z is closed under addition and have *verified* that it is closed under multiplication.

Regarding our candidate system, $[D,+,\bullet]_A$, we have shown that it may be the system $[Z,+,\bullet]_A$. It all depends on whether Z can be proved closed under multiplication. Nevertheless, we will re-label our system's nomenclature from $[D,+,\bullet]_A$ to $[Z,+,\bullet]_A$ to express an appreciation of the extreme likelihood that Z is closed under multiplication. Of course, because all of the elements of Z *are* numbers, the system categorically conforms to all of the axioms. The systems $[E_{0-},+,\times]$, $[N_{0-},+,\times]$ and $[C,\oplus,\otimes]$, however, require some further investigation.

The system $[E_{0-},+,\times]$ fails to meet the criteria of this axiom because the set E_{0-} is composed of just the even numbers (positive and negative) and zero and does not have either the multiplicative identity element 1 or the additive inverse of 1. All total, it is missing the following numbers

$$\ldots\ -7,\ -5,\ -3,\ -1,\ 1,\ 3,\ 5,\ 7,\ \ldots,$$

which can be constructed from these two numbers, and which the closure axiom of addition demands it must have.

If we include these numbers in the set, then we will just have the set of integers, Z, all over again. Consequently, we can do away with the contrived system $[E_{0-},+,\times]$ and just consider the axiom-given system $[Z,+,\bullet]_A$. Furthermore, because the system $[N_{0-},+,\times]$ is already identical to the system $[Z,+,\bullet]_A$, for all intents and purposes, but was also contrived, we can eliminate it as well.

Examining the system $[C,\oplus,\otimes]$ and its multiplication table as shown in Table 6 (section 4.1), we observe that the set C does have a multiplicative identity element, which happens to be, in this case, the number 1 itself. Obviously, if we wrap-around multiply any number from the set C by the number $1 \in C$, then the result will be identical to the number that was multiplied.

We are now down to only two viable systems for algebra: $[Z,+,\cdot]_A$ and $[C,\oplus,\otimes]$.

~

4.5 Multiplicative Inverses and Division

We have all been schooled in fractions. Fractions, the very word may conjure up some of your deepest fears. Dreams of parading in shabby underpants or of having your teeth fall out like white shattered stone may have been more desirable than finding the least common denominator on a high school test. You may give thanks to the ancient art of commerce (because the ability to calculate the sale price of a third of a loaf of bread was essential to making money) and the next axiom for the onset of your anxieties.

It, like a few of the axioms before it, is recursive. It calls into existence, a plethora of new numbers, called multiplicative inverses, and multiplicative inverses of multiplicative inverses, and ... well, you can infer the rest.

However, this axiom, like most of the axioms before it, is dependent upon the existence of some numbers in the first place, and, as it just so happens, the axioms have just given birth to the integers. Thus, according to the next axiom, every one of those integers (except zero) has a multiplicative inverse, but we will discuss that matter in greater detail a little further on.

❖ **Axiom M5: The Existence of Multiplicative Inverses**

*For every number a, $a \neq 0$, there is a number called the multiplicative inverse of a, written as $\dfrac{1}{a}$ (over-under notation), a^{-1} (exponential notation) or $1/a$ (side-over-side notation), such that $a \cdot \dfrac{1}{a} = 1$, $a \cdot a^{-1} = 1$ or $a \cdot (1/a) = 1$**

Many of you are familiar with the use of exponents as the notation to describe raising a number to a power. However, for the time being the negative-one exponent being used in this axiom is to be understood strictly as the notation for indicating the multiplicative inverse of a number.

On various occasions, we will be using either the notation a^{-1} (exponential) or the notation $1/a$ (side-over-side) instead of the preferred over-under notation whenever we write multiplicative inverses in line with the text because it makes the text more comfortable to read.

However, exponential and side-over-side notation has its limitations and can be difficult to understand, especially when talking about rational numbers. The placement of the parentheses must be strictly observed whenever using either exponential or side-over-side notation; therefore, we will quite often

---

* Multiplicative inverses are unique, again I leave it to you to show.

revert to the more familiar over-under notation.

To avoid making communication errors, because of improperly placed parentheses, think about what it is you want to say before you write. (I can only pray upon raw knees that I have done the same throughout the course of this book.) For example, suppose a and b are numbers, then by Axiom M1 ab is a number, and because ab is a number, M5 states it has an inverse, which is $(ab)^{-1}$. The important point being made here is that we are considering ab as *one* number. It would be semantically incorrect to write the multiplicative inverse of ab as $a^{-1}b^{-1}$, even though the two expressions $(ab)^{-1}$ and $a^{-1}b^{-1}$ are equal (see Theorem 4.5.3). If we were to put each into words, then $(ab)^{-1}$ reads as *the multiplicative inverse of the product of* a *and* b; while $a^{-1}b^{-1}$ reads as *the product of the multiplicative inverses of* a *and* b. The presence and location of the parentheses in the former number results in a completely different meaning as compared to the latter number.

Exponential notation for multiplicative inverses is a widely-used notation, and it behooves one to become familiar with it regardless of its drawbacks. If you have any trouble making sense of the notation a^{-1}, then I urge you to convert it to fractional notation whenever it occurs until you are totally familiar it. For your reference, I have included the following notational equalities. Each has its pros and cons.

$$\frac{1}{a\cdot b}=(a\cdot b)^{-1}=1/(a\cdot b)\text{, whereas }\frac{1}{a}\cdot\frac{1}{b}=a^{-1}\cdot b^{-1}=(1/a)\cdot(1/b),$$

$$\text{and }\frac{1}{\frac{1}{a}}=(a^{-1})^{-1}=1/(1/a)\text{ (where neither } a \text{ nor } b \text{ equal zero).}$$

Three essential points (theorems) remain that need to be dispensed with before we can move on to other matters regarding this axiom and the order of the integers.

Point 1: $1=1^{-1}$; in other words, *the number* 1 *is equal to its multiplicative inverse*. The proof is analogous to that of $0=-0$. I leave it for you to prove (and, please, cut and paste in your mind's eye as you like from Theorem 3.5.1).

Point 2: $(a^{-1})^{-1}=a$. Likewise, the proof of this statement is analogous to the proof that the additive inverse of the additive inverse of a number is the number itself (i.e., it is analogous to the proof that $-(-a)=a$).

If you would like to prove this about multiplicative inverses on your own, then do not read the proof of the next theorem because it has been literally copied from the proof for Theorem 3.5.2, and then pasted below to expressly show the analogous structures of the addition axioms and the multiplication axioms. However, the theorem is important and useful on its own.

Throughout the proof, I have exchanged the word *add* with the word *multiply*, the operator $+$ with the operator of $\cdot$, the number 0 with the number 1, and $-a$ with $1/a$. Please take a moment to compare the two proofs.

This theorem states that *the multiplicative inverse of the multiplicative inverse of a number is the number.*

◆ **THEOREM 4.5.1**

$$\frac{1}{\left(\frac{1}{a}\right)}=(a^{-1})^{-1}=a,\ a\neq 0$$

Proof: a^{-1} is a number by M5 (that is, if $a\neq 0$ is a number, which we assume it is), and it therefore has a multiplicative inverse (by this axiom), which is, accordingly, $(a^{-1})^{-1}$ and it is such that $(a^{-1})\cdot(a^{-1})^{-1}=1$. Now, the multiplication property of equality (Chapter 2, section 2.1) allows us to multiply both sides of this equality by the number a; hence, $a\cdot[(a^{-1})\cdot(a^{-1})^{-1}]=a\cdot 1$. The associative property, M3, allows us to shift the square brackets on the left hand side of this equality giving us $[a\cdot(a^{-1})]\cdot(a^{-1})^{-1}=a\cdot 1$. However, a^{-1} is the multiplicative inverse of a according to M5; hence the product of these two numbers is equal to 1. Therefore, we have $1\cdot(a^{-1})^{-1}=a\cdot 1$, and 1 multiplied by any number is the number according to M4. Consequently, we have the desired result:

$$\frac{1}{\left(\frac{1}{a}\right)}=(a^{-1})^{-1}=a.$$

In short a and a^{-1} are multiplicative inverses of each other. From a more practical point of view, we can now say, for instance, that

$$\frac{1}{\frac{1}{5}}=\frac{1}{\left(\frac{1}{5}\right)}=(5^{-1})^{-1}=5.$$

[It is crucial to recognize the lengths or thicknesses of the division lines in the above expression because that is what communicates precisely the way numbers divide each other. (The division line—for over-under notation only—is a substitute for parentheses.) Ergo, there should be no confusion regarding the order of operations with this notation.]

Point 3: $-1<0<1$. According to Axiom M5, every number, *except* 0, has a multiplicative inverse. In other words, there is *no* number that we can multiply 0 by, such that the result will equal 1. In fact, 0 is the only number on which any exception to the axioms is imposed at all, otherwise 0 revels in all of the "rights" and "privileges" of the axioms like every other number.

Because of this exception, however, we can prove by contradiction (Section 1.2) that $0\neq 1$. And it must be proved because we are dealing with *symbols* that need to be ordered, for no reason other than to see where our logic takes us and to coincide with what we know about numbers in our everyday lives.

Proof: Suppose $0=1$, then by the multiplication property of equality (E5), we can multiply both sides of this equality by the number 1 and obtain $1\cdot 0=1\cdot 1$. Axiom M4, however, states that $1\cdot 1=1$, then by the transitive property of equality (E3), we have $1\cdot 0=1$. This last equality implies that there *is* a number that we can multiply 0 by to get 1, contrary to axiom M5, which states that there is no such number. Hence it cannot be the case that $0=1$, therefore it must be the case that $0\neq 1$.* Axioms, like the Ten Commandments, may be neither broken nor denied.

We have been writing the set of integers, Z, in an orderly fashion. However, nothing has been said (so far) that would lead us to suspect that $0<1$, $1+1<1+1+1$, $-5<2$, or that $-1<0$. To discover how these symbols are ordered, we must apply logic. We have already shown that $0\neq 1$ and by the law of trichotomy (O1), therefore, it must be the case that either $0<1$ or $1<0$ (but not both). If we can show that it is not one of these two, then it must be the other. We will prove that $1<0$ is false, by using, once again, the method of contradiction, and therefore conclude that $0<1$ is the correct order.

Proof: Suppose $1<0$, then by the addition axiom for inequalities (O3), we may write $1+(-1)<0+(-1)$. However, $1+(-1)=0$ because the sum of additive inverses is 0 (A5), and $0+(-1)=-1$ because that is a property of 0 (A4). Making the appropriate substitutions (substitution property of equality (E6)), we may write $0<-1$. Having met all of the conditions of the multiplication property of inequalities (Axiom O4), we are now able to use it. It states,

If $a<b$ *and* $0<c$, *then* $a\cdot c<b\cdot c$.

Consequently, if $1<0$ (which we have supposed it is) and $0<-1$ (which it would have to be given that supposition), then $1\cdot(-1)<0\cdot(-1)$ by O4. However, on the left side of $<$ we have $1\cdot(-1)=-1$, because that is a property of the number 1 (A4); while on the right side we have $0\cdot(-1)=0$ because that is a property of the number 0 by Assumption #1 (we have a lot riding on this assumption). Making the appropriate substitutions (substitution property of equality, E6), we obtain $-1<0$. (This last inequality could, itself, be used to show a contradiction, but I am choosing to trudge onward to bring the point home.) However, if we add 1 to both sides of this inequality (addition property of inequalities), we obtain $-1+1<0+1$, but that means that $0<1$. However, this contradicts our original supposition that $1<0$. Therefore, both $1<0$ is false, and $0=1$ is false. Consequently, by the law of trichotomy, it must be the case that $0<1$.† However, $0<1$ implies $-1<0$. Therefore, by virtue of the transitive property of inequalities (O2), we have proved the point that we set out to prove, and that is that $-1<0<1$.

---

* This is the Law of the Excluded Middle: either a proposition is true, or its negation is true.

† If we were to try and prove that $0 < 1$ by assuming that $0 < 1$, then we would have run into a tautology, which would be useless. In other words, proving $0 < 1$ by assuming it is useless.

There is, however, one heck of a wild bull sneaking about in the fine china and crystal shop that needs to be addressed. What if it is the axioms themselves which contradict each other? How can we be sure they do not? How can we be sure that from among all the logical statements being deduced and written that somewhere down the line we won't arrive at a statement that we know in our heart of hearts to be true, but which the axioms by way of deduction say is false, or the other way around, or for which the axioms cannot decide one way or the other because, perhaps, of some missing axiom that we have failed to include? The answer is that we cannot be sure.

Although many mathematicians have examined the axioms for flaws, contradictions and inclusiveness in the hopes of mustering and proving everything that could be said about algebra until there was nothing left unsaid and unknown about algebra, it has been proven to be a logical and mathematical fantasy—recall Whitehead and Russell and their *Principia Mathematica* (Section 1.2). However, the axioms of algebra have been vetted repeatedly by just as many mathematicians for well over a hundred years. Consequently, we have little choice but to accept the axioms as the supreme law of the land. Nothing better has been found to replace them.

Although the proof that $0<1$ did not depend upon any interpretation of the symbol $<$, this symbol has the historical meaning of *less than*. We are now able to formally order the set $Z=\{\ \ldots-3,\ -2,\ -1,\ 0,\ 1,\ 2,\ 3,\ \ldots\ \}$, which hitherto has only been ordered for the benefit of your comprehension.

Because we have shown that $0<1$ and because we may add 1 to both sides of this inequality by Axiom O3 and to both sides of the inequalities produced thereafter, we may write

$$1<1+1,\quad 1+1<1+1+1,\quad 1+1+1<1+1+1+1,\ \text{etc.},$$

by definition (4.4.1), but these sequences of 1's and plusses may be written as

$$1<2,\quad 2<3,\quad 3<4,\ \text{etc.}$$

Furthermore, if we continuously add -1 to both sides of $-1<0$, we obtain

$$-1+(-1)<-1,\quad -1+(-1)+(-1)<-1+(-1),\quad -1+(-1)+(-1)+(-1)<-1+(-1)+(-1)\ldots$$

And, by definition (4.4.2), each of these are just

$$-2<-1,\quad -3<-2,\quad -4<-3,\ \text{etc.}$$

Finally, with the help of (multiple counts of) the transitive property of inequalities (O2), the order of the integers from lesser to greater has been established. It is what we have always known it to be:

$$\ldots-4<-3<-2<-1<0<1<2<3<4\ldots$$

Thus, we can say in an absolute way (as opposed to the relative way that was discussed in the introduction to section 3.5) that $-9<-1$, $-3<2$, and $7<9$.

Although we may have logically ordered the integers, which is certainly comforting, this new axiom has caused the creation of infinitely many more numbers, and they need to be sorted and ordered as well.

On first impression, they are just the set of the multiplicative inverses of the integers (except 0), which I will occasionally call *unit fractions* (i.e., numbers that have 1 in the numerator and an integer (not 0) in the denominator). We will denote the set of unit fractions (multiplicative inverses of the integers) by

$$Z^{-1}=\left\{\frac{1}{-1},\ \frac{1}{-1+(-1)},\ \frac{1}{-1+(-1)+(-1)},\ \cdots\ \cdots\ \frac{1}{1+1+1},\ \frac{1}{1+1},\ \frac{1}{1}\right\}.$$

Because we have already defined the numbers in the denominators as integers, we may substitute (E6) and rewrite these numbers as follows (again, the order for the moment is only assumed):

$$Z^{-1}=\left\{\frac{1}{-1},\ \frac{1}{-2},\ \frac{1}{-3},\ \cdots\ \frac{1}{3},\ \frac{1}{2},\ \frac{1}{1}\right\}=\left\{(-1)^{-1},(-2)^{-1},(-3)^{-1}\ldots\ 3^{-1},2^{-1},1^{-1}\right\}. \quad \blacksquare 4.5.1$$

Fortunately, once we prove that the product of any pair of integers is an integer, then we will have also proved that the product of any pair of numbers from the set Z^{-1} is also a number in the set Z^{-1}, which only depends on how unit fractions are multiplied (which you already know $(1/a)(1/b)=1/(ab)$).

However, before we can formally show how to do that, we must first prove a rather small but important theorem that states if each of two numbers is not equal to zero, then their product is also not equal to zero.

This may sound strange to your ears because you have never experienced a single case where the product of any two non-zero numbers resulted in zero. That does not prove, however, that such a pair of non-zero numbers does not exist. Although we intuitively believe in our heart of hearts that such a pair of numbers could not possibly exist, logically we cannot be certain because we are dealing with "meaningless" symbols that are grounded in axioms, thus we are obliged to prove it; the proof is by contradiction.

◆ THEOREM 4.5.2 AND ITS COROLLARY

Theorem: *If* $a\neq 0$ *and* $b\neq 0$ *are numbers, then* $ab\neq 0$

Corollary: *If* $ab=0$*, then either* $a=0$ *or* $b=0$*, (or both)*

Proof of theorem: Suppose the contrary, that $ab=0$, then because $a\neq 0$ and $b\neq 0$, these two inequalities have multiplicative inverses, and multiplying both sides of $ab=0$ by the number a^{-1} (multiplication axiom of equality), we obtain $a^{-1}(ab)=a^{-1}\cdot 0$, but by the associative axiom of multiplication we have $(a^{-1}a)b=a^{-1}\cdot 0$. However, the product of multiplicative inverses is 1, and by Assumption #1, any number multiplied by 0 is 0; so we are left with $b=0$, but this contradicts our premise that $b\neq 0$. Because of the symmetrical properties of the word *and,* we could have just as well used the multiplicative inverse of

b in the proof and concluded that $a=0$, which is also a contradiction. Apparently, one or the other of the two conditions of our premise must be zero if $ab=0$. Therefore, our assumption that $ab=0$ is false, and consequently $ab\neq 0$. The corollary is the contrapositive of the theorem; hence, it too is true.

Given Theorem 4.5.2, we can now prove the next theorem.

❖ THEOREM 4.5.3

For numbers $a\neq 0$ *and* $b\neq 0$, $\frac{1}{a}\cdot\frac{1}{b}=\frac{1}{a\cdot b}$

Proof: if a and b are numbers, neither of which are 0, then ab is a number which is not 0 by Theorem 4.5.2. Hence all three of these numbers have multiplicative inverses and we may write

$$(ab)\cdot\frac{1}{ab}=1$$

Multiplying both sides of this equation by the product of the multiplicative inverses of b and a, we obtain

$$\left(\frac{1}{b}\cdot\frac{1}{a}\right)\cdot\left((a\cdot b)\cdot\frac{1}{a\cdot b}\right)=\left(\frac{1}{b}\cdot\frac{1}{a}\right)\cdot 1 \qquad \blacksquare 4.5.2$$

On the *left side* of this equality, after some shifting of the parentheses and employing the properties of multiplicative inverses and of 1, we obtain

$$\frac{1}{b}\cdot\left(\frac{1}{a}\cdot a\right)\cdot b\cdot\frac{1}{a\cdot b}=\frac{1}{b}\cdot(1)\cdot b\cdot\frac{1}{a\cdot b}=\left(\frac{1}{b}\cdot b\right)\cdot\frac{1}{a\cdot b}=(1)\cdot\frac{1}{a\cdot b}=\frac{1}{a\cdot b},$$

to be clear, on the left side of the equality shown in 4.5.2 we obtain

$$\frac{1}{a\cdot b}.$$

While on the *right side* of that equality we have

$$\left(\frac{1}{b}\cdot\frac{1}{a}\right)\cdot 1=\frac{1}{b}\cdot\frac{1}{a}.$$

Because these two sides are equal, we have, after applying M2 and implementing the symmetric property of equality, the following equality:

$$\frac{1}{a}\cdot\frac{1}{b}=\frac{1}{a\cdot b}.$$

In words, we would say that *the product of the multiplicative inverses of two numbers is equal to the multiplicative inverse of the product of the two numbers.* A practical example of this proof using a pair of numbers from Z^{-1} is

$$\frac{1}{3}\times\frac{1}{5}=\frac{1}{3\times 5}. \qquad \blacksquare 4.5.3$$

Because the numbers in the denominators to the left of $=$ are integers, it will follow conclusively that if the product of any pair of integers is an integer, then the product of any pair of unit fractions is a unit fraction; in other words, the product will be the multiplicative inverse of an integer.

However, the closure axioms also demand that we determine what the sum and product of any number from Z and any number from Z^{-1} is. These numbers originated—emerged—from the axioms, and we must see where the closure axioms ferry them and the constructions they produce whenever any pair of such numbers are either added or multiplied.

We will denote the set that contains the *products* of every possible pairing of the numbers from each of the sets Z and Z^{-1}, by the set $Z \otimes Z^{-1}$.* Thus, if $x \in Z$ and $1/y \in Z^{-1}$, then $x \cdot (1/y) \in Z \otimes Z^{-1}$ (Does the set that contains the *addition* of every possible pairing [i.e., $Z \oplus Z^{-1}$] equal $Z \otimes Z^{-1}$? ☆8)

Normally, this process of creating a set would not guarantee closure under multiplication or addition because we would still have to consider the products of products and sums of sums and include them in the set, and then take the products and sums of those numbers and include them in the set, and so on, akin to creating the closed set from the finite set $\{0, 3, 5\}$ as was done in Section 3.1.

As a verification of this, suppose we are given the set of every prime number ending in either 1 or 3 and the set of every prime number ending in ether 7 or 9 as shown immediately below:

$$P_{13} = \{3, 11, 13, 23, 31, 41, 43, \ldots\}, \quad P_{79} = \{7, 17, 19, 29, 37, 47, \ldots\}.$$

Then the set of the products of all possible pairings of the numbers from these two sets (i.e., the set of all $xy \in P_{13} \otimes P_{79}$, such that $x \in P_{13}$ and $y \in P_{79}$) is the set $P_{13} \otimes P_{79} = \{3\times 7,\ 3\times 17,\ 3\times 19,\ \ldots\ 11\times 7,\ 11\times 17,\ 11\times 19,\ \ldots\ 13\times 7,\ 13\times 17,\ \ldots\}$, which equals $\{21, 51, 57, \ldots\ 77, 187, 209, \ldots\ 91, 221, \ldots\}$. This set, however, is not closed under multiplication or addition because, case in point, although the numbers 21 and 77 are in the set, the numbers 21×77 and $21+77$ are not. In fact, the product or sum of any two numbers from the set $P_{13} \otimes P_{79}$ is not in the set $P_{13} \otimes P_{79}$. This means, by definition, that the set $P_{13} \otimes P_{79}$ is not closed under either operation, and the only way to close it would be to include every pairwise product and sum of every pairwise product and sum of every pairwise product and sum … ad infinitum of all these numbers in the set. To be clear, however, that is not what the operator $\otimes$ between two sets does. It only takes the elements of one set and pairs each of them with every element of another set and creates the set of all product pairs. *Period.*

However, the set $Z \otimes Z^{-1}$ is different from this and most other sets. It *is* closed under multiplication and addition and we will prove it in the next section. Furthermore, if we can show that the addition of any two numbers from the sets Z and Z^{-1} is a number in the set $Z \otimes Z^{-1}$ (i.e., $Z \oplus Z^{-1} \subseteq Z \otimes Z^{-1}$, where $\oplus$ does what $\otimes$ does, but for addition), then we will have not only found the

* For those of you interested in how such a set is constructed, I refer you to Appendix B.

smallest set of numbers that satisfies all of the conditions of every axiom, but we will have also found the set of numbers that is completely derivable from the axioms. (I refer you to Appendix B to see how the set $Z \otimes Z^{-1}$ is created.)

However, before we can consider the product and sum of any of the numbers in the set $Z \otimes Z^{-1}$, we still need to determine what type of number the product of an integer and the multiplicative inverse of an integer is. Is it an integer or the multiplicative inverse of an integer? It turns out that it is neither of these two options; however, we will define the product in such a way to create a new class of numbers.

Currently, however, the best that we can do with the product of an integer and the multiplicative inverse of an integer is to write it as, well, the product of an integer and the multiplicative inverse of an integer. It cannot be expressed in any other way. Some examples of what I mean are shown below:

$$11 \times \frac{1}{21},\ (-2) \times \frac{1}{5} \text{ or more generally } a \cdot \frac{1}{b}\ \ (b \neq 0).$$

These numbers *cannot*, at least for the moment, be written in any other way.

Now, many of you may find this assertion difficult to accept, and be immediately inclined to put forth the answers to the above products as

$$\frac{11}{21},\ \frac{-2}{5} \text{ or more generally } \frac{a}{b}\ \ (b \neq 0). \qquad \blacksquare 4.5.4$$

Although you would be, academically, correct to do so, there is absolutely no *axiom* or *theorem* of algebra that will allow us to write this configuration of symbols;* that is, we cannot write a configuration where a number other than 1 is in the numerator. The reason is that the individual elements that are used to make up the entire symbol for multiplicative inverses are considered inseparable. Hence, there must *always* be a number 1 above the division line, and the numeral-number in question, whose multiplicative inverse is being displayed, must *always* appear below the division line; that is how multiplicative inverses are written—that is what Axiom M5 states. Thus, technically, we can never formally write any of the numbers as portrayed in 4.5.4.

Furthermore, if one wanted to write a number like

$$\frac{\frac{2}{5}}{\frac{3}{7}},$$

which has always been a perfectly fine presentation of a fractional number—two-fifths divided by three-sevenths—then the best that one could do with the axioms, theorems and definitions on hand is to write it as a mix of integers and unit fractions:

---

* The same was true about subtraction: "$a-b$" is impossible to deduce without a definition.

$2\times\frac{1}{5}\times\frac{1}{3\times\frac{1}{7}}$ (which would equal $14\times\frac{1}{15}$, and *not* $\frac{14}{15}$ as you might expect).

However, this portrayal of the numbers is difficult to work with to say the least, even if we were to introduce parentheses (recall that the division line replaces the need to use parentheses for, *only*, over-under notation). In exponential form, it would be just as difficult to understand: $2\times5^{-1}\times(3\times(7^{-1}))^{-1}$.

The gist of this is that because the product of an integer and the multiplicative inverse of an integer cannot be written in any other way, these products cannot be classified as either integers or the multiplicative inverses of integers (although some products are either). They represent a class of numbers called the *rational numbers,* which we will define presently. Regardless of what we choose to call these products, this is such a cumbersome way to write such a pair of these factors that it quickly becomes annoying.

To get around this little hiccup, mathematicians have devised the next definition, which is often thought of as division, but which really depends on one's perspective. Often, it is just thought of as an abbreviation for multiplication or simply a number. Like subtraction, it *forces* the equality between two different configurations of symbols.

∘ DEF. 4.5.1: DIVISION AND QUOTIENTS

For numbers a and b $(b\neq0)$, we define the quotient of a and (or with, or divided by) b as $\frac{a}{b}=a\cdot\frac{1}{b}$.

With this definition, we now have a logical justification for writing the numbers shown in line 4.5.4, and a method for multiplying pairs of fractions to form a single fraction. For example,

$$\underset{\text{Given}}{\frac{2}{3}\times\frac{7}{6}} = \underset{\text{Definition 4.5.1}}{\left(2\times\frac{1}{3}\right)\cdot\left(7\times\frac{1}{6}\right)} = \underset{\text{Commute/Associate}}{(2\times7)\times\left(\frac{1}{3}\times\frac{1}{6}\right)} = \underset{\text{Theorem 4.5.3}}{(2\times7)\times\left(\frac{1}{3\times6}\right)} = \underset{\text{Definition 4.5.1}}{\left(\frac{2\times7}{3\times6}\right)},$$

which is just the method we have been taught for multiplying fractions: multiply across the numerators and across the denominators to obtain the product.

However, a definition does not change the logic behind the definition. The fact of the matter is that the product of an integer and the multiplicative inverse of an integer is still just that—it doesn't get any better. However, the definition of division and quotients, beckons us forward to the next definition.

∘ DEF. 4.5.2: RATIONAL NUMBERS

A rational number is any number that can be expressed as the quotient of two integers (0 not in the denominator). The set of all rational numbers is represented by Q.

Now, by Def. 4.5.1 (Division) and because of the manner in which the set $Z\otimes Z^{-1}$ was defined, it is clear that every number in the set $Z\otimes Z^{-1}$ is a number in the set Q, i.e., $Z \otimes Z^{-1} \subseteq Q$ (because Q is the set of *all* rational numbers and there is no number in $Z \otimes Z^{-1}$ that cannot be *expressed* as the quotient of two integers). Furthermore, because the set Q is composed of the quotient of every pair of integers (0 not in the denominator), and there are no other integers except for those defined in the set Z and in the denominators of the numbers in set Z^{-1}, it follows that $Q \subseteq Z \otimes Z^{-1}$. It must be the case, therefore, that $Z \otimes Z^{-1} = Q$.*

However, we will withhold making this claim official until after we show that the product of two integers is an integer and the sum of any pair of numbers from the set $Z \otimes Z^{-1}$ is a number in the set $Z \otimes Z^{-1}$.

In the meantime, we can remove any repeats from the set Q and from the set $Z \otimes Z^{-1}$ (i.e., fractional equivalents such as 2/4, which is equivalent to 1/2, or 3/9, which is equivalent to 1/3, etc.) because if the numerator and the denominator have any common factors, then we can "eliminate" or cancel them. For example, in the fraction below the common factor c was canceled.

$$\frac{ac}{bc} = \left(\frac{a}{b}\right)\left(\frac{c}{c}\right) = \frac{a}{b}, \text{ where } b \neq 0 \text{ and } c \neq 0.†$$

There is no easier way to display the set of rational numbers than in a table (dashes represent repeats).

$Q =$

$0/1$	$1/1$	$-1/1$	$2/1$	$-2/1$	$3/1$	$-3/1$	$4/1$	$\cdots$
	$1/2$	$-1/2$	-	-	$3/2$	$-3/2$	-	$\cdots$
	$1/3$	$-1/3$	$2/3$	$-2/3$	-	-	$4/3$	$\cdots$
	$1/4$	$-1/4$	-	-	$3/4$	$-3/4$	-	$\cdots$
	$1/5$	$-1/5$	$2/5$	$-2/5$	$3/5$	$-3/5$	$4/5$	$\cdots$
	$\vdots$	$\vdots$	$\vdots$	$\vdots$	$\vdots$	$\vdots$	$\vdots$	$\ddots$

Table-8: The Set of Rational Numbers (Q stands for "Quotient")

There is one more important and frequently needed theorem that we must prove in this section before examining our contenders for the game of algebra and before finally moving on to the last section of this chapter.

In words, it states that *the additive inverse of the quotient of* a *with* b *is equal to the quotient of the additive inverse of* a *with* b*, which is equal to the quotient of* a *with the additive inverse of* b. How magnificently easy is all of this said in the language of mathematics (overleaf):

---

* This is an example of an intuitive "proof", which—my apologies—is also rather subtle.

† To avoid this slightly messy bit about equivalent fractions, mathematicians use an idea called equivalent classes to construct rational numbers. Equivalent classes partition sets into groups of equal elements. We have seen examples of equivalent classes when we partitioned the set of numbers we obtained from the closure axioms into three groups.

◆ **THEOREM 4.5.4**

For numbers a, and $b \neq 0$, $-\dfrac{a}{b}=\dfrac{-a}{b}=\dfrac{a}{-b}$

Knowledge of the ability to move the dash (the negative symbol) from among the middle, the numerator and the denominator of a quotient is essential for proficiency in algebra.

We can show the theorem is true rather quickly, but I will leave it for you to supply the missing parts and reasons for the equalities shown below.

$$\underset{\text{Given}}{-\frac{a}{b}} = \underset{?}{(-1)\left(a\cdot\frac{1}{b}\right)} = \underset{?}{(-1\cdot a)\left(\frac{1}{b}\right)} = \underset{\text{Def. 4.5.1}}{\left(\frac{-1\cdot a}{b}\right)} = \underset{?}{\frac{-a}{b}} = \underset{?}{\frac{-1}{-1}\cdot\frac{-a}{b}} = \underset{?}{\frac{-1\cdot(-a)}{-1\cdot b}} = \underset{?}{\frac{a}{-b}}. \quad ■4.5.6$$

Regarding our two remaining candidate systems, the discovery system should be upgraded, owing to our discussion in this section, to $[Z\otimes Z^{-1},+,\bullet]_A$.

However, looking at the wrap-around multiplication table (Table 6) for the number 12 in section 4.1, the system $[C,\oplus,\otimes]$ only has multiplicative inverses for the numbers 1, 5, 7 and 11: $1\otimes1=1$, $5\otimes5=1$, $7\otimes7=1$ and $11\otimes11=1$.* The numbers 2, 3, 4, 6, 8, 9 and 10 do not have multiplicative inverses. The number 12 was our zero, hence, we can simply stipulate (define) that it does not have one.

We could include the missing inverses in the set of clock numbers, for example, for the number 9 we could include the number $1/9$, but then our clock would have thirteen numbers in it. Furthermore, the closure property of addition would demand that the number $1/9+1/9=2/9$ also be included in the set of clock numbers.

Our clock, however, was supposed to only have twelve numbers in it, and it would have many more numbers if we were to include the multiplicative inverses of all the clock numbers that do not have one, not to mention the sums and products of these numbers. The system $[C,\oplus,\otimes]$ would mushroom wildly beyond bounds and that was clearly not what was intended for the set C, and it is difficult to see how we can define our way out of it; so, we will let it go. Consequently, the only system that we have left is $[Z\otimes Z^{-1},+,\bullet]_A$, the one given to us by the axioms themselves.

That does not mean there are no contrived sets or systems with oddly defined additions and multiplications that will conform to the axioms, there are. It is just that our contrived systems did not.

This was a rather long section, however, if you have managed to make it this far, then you cannot give up now. We have finally made it to the long-awaited distributive axiom and the end of this chapter.

~

---

* Notice that $1^{-1}=1$, $5^{-1}=5$, $7^{-1}=7$, and $11^{-1}=11$.

4.6 THE DISTRIBUTIVE PROPERTY

We have waited long enough, so let's get to it.

❖ AXIOM D: DISTRIBUTION

For numbers a, b *and* c, $c(a+b)=ca+cb$; *hence,* $(a+b)c=ac+bc$ *(by* M2 *&* E3*)*

Thus technically, there is no such thing as "adding like terms", there is only the distributive axiom: $3x+5x=(3+5)x=8x$ or $7x^2y+10x^2y=(7+10)x^2y=17x^2y$.

As promised, we will first prove (two methods) that zero times any number is zero, which was the statement of Assumption #1.

◆ THEOREM 4.6.1

For every number a, $a\cdot 0=0$

Proof (method 1): $0+0=0$ (property of zero), so $a(0+0)=a\cdot 0$ (multiplication property of equality), hence, $a\cdot 0+a\cdot 0=a\cdot 0$ (Distributive axiom). Because $a\cdot 0$ is a number (A1), and all numbers have additive inverses, we can add $-(a\cdot 0)$ to both sides using the addition property of equality to obtain $(a\cdot 0+a\cdot 0)+(-a\cdot 0)=a\cdot 0+(-a\cdot 0)$. Rearranging the terms and the parentheses (A2 and A3) and employing the property of additive inverses and the property of zero (A5 and A4), we have the desired result: $a\cdot 0=0$. In particular, $0\cdot 0=0$.

Proof (method 2): $0+0=0$ (property of zero), so $a(0+0)=a\cdot 0$ (multiplication property of equality), hence, $a\cdot 0+a\cdot 0=a\cdot 0$ (Distributive axiom). However, only one number has the property to return the identity of any number added to it, and that number is 0. Consequently, it must be the case that $a\cdot 0=0$.

Second, we will prove assumption #2, which stated that *the product of* (-1) *and any number is the additive inverse of the number.*

◆ THEOREM 4.6.2

For every number a, $(-1)\cdot a=-a$

Proof: $1+(-1)=0$ (Axiom A5), $(1+(-1))\cdot a=0\cdot a$ (Axiom E5), $1\cdot a+(-1)\cdot a=0$ (Axiom D, and Theorem 4.6.1), and $(a+(-1)\cdot a)+(-a)=0+(-a)$ (property of 1 and addition property of equality); hence, $(a+(-a))+(-1)\cdot a=0+(-a)$ after re-arranging terms and shifting parentheses; finally, using the properties of additive inverses and 0, we have $(-1)\cdot a=-a$. In particular, $(-1)(-1)=-(-1)$ by this theorem and Theorem 2.5.2. Consequently, every logically valid argument that used either of these assumptions is now logically sound.

We can extend the distributive axiom in many ways. For example, we can show that if v, w, x, y and z, are numbers, then

$$\underbrace{(v+w)}_{\text{2 terms}}\cdot\underbrace{(x+y+z)}_{\text{3 terms}}=\underbrace{vx+vy+vz+wx+wy+wz}_{\text{6 terms}}. \qquad ▪4.6.1$$

The reason is that $(v+w)$ in 4.6.1 can logically be thought of as one number; hence, we may replace it with the number c that was exhibited in Axiom D.

Furthermore, $(x+y+z)$ in 4.6.1 can be thought of as the sum of the two numbers $(x+y)$ and z; each of which can be used to replace the numbers a and b exhibited in that axiom, respectively; thus, mimicking the form required by that axiom; hence,

$$(v+w)\cdot(x+y+z)=\underbrace{(v+w)}_{c}\underbrace{\cdot}_{\cdot}\underbrace{((x+y)+z)}_{(a+b)}\underbrace{=}_{=}\underbrace{(v+w)}_{c}\underbrace{\cdot}_{\cdot}\underbrace{(x+y)}_{a}\underbrace{+}_{+}\underbrace{(v+w)}_{c}\underbrace{\cdot}_{\cdot}\underbrace{(z)}_{b}=$$

$$(v+w)x+(v+w)y+(v+w)z=vx+wx+vy+wy+vz+wz. \qquad \blacksquare 4.6.2$$

After rearranging the terms in 4.6.2, we obtain the equation shown in 4.6.1.

Mechanical proficiency in algebra is, to some degree, about finding patterns. Notice that the v in the first factor of the first expression in 4.6.1 was multiplicatively distributed to each of the numbers in the second factor, and the same was done for the w in that first factor, after which they were all summed together. We could continue this pattern for any number of factors, each composed of any number of numbers.*

We can now demonstrate that the product of any two integers is itself an integer; in doing so, we will have demonstrated that the set Z, and therefore the set Z^{-1} are closed sets under multiplication. With that information on hand, we will also be able to demonstrate that the set $Z\otimes Z^{-1}$ is a closed set under both multiplication and addition.

To prove that the product of two integers is an integer, we will begin by supposing that the first integer factor consists of any number of terms, say j terms, each of which, we will stipulate, is entirely the number 1, or entirely the number -1. We will label this first integer as A. Thus, A is either a positive or a negative integer. The second factor will also consist of any number of terms, say k terms, each of which, we will stipulate, is *only* the positive number 1. The result of this set-up is the following two-factor product:

$$A\cdot(\underbrace{1+1+1+\ \dots\ +1}_{k \text{ number of 1s}})$$

Based upon the distributive axiom and our previous discussion about extending this axiom, we may distribute the factor A to each of the 1s in the second factor multiplicatively, which will give us k terms, each of which consists of the product $A\cdot 1=A$, hence we have the following result:

$$\underbrace{A\cdot 1+A\cdot 1+A\cdot 1+\ \dots\ +A\cdot 1}_{k \text{ number of } A\cdot 1\text{s}}=\underbrace{A+A+A+\ \dots\ +A}_{k \text{ number of } A\text{s}}.$$

Because we have already shown that the sum of any two integers is an integer, and A is an integer, it follows that the expression on the right of equals in the above equation is either a positive integer or a negative integer depending upon the composition of A. The actual value of the integer, which in this case

*In many ways, this is how the set $Z\oplus Z^{-1}$ was constructed, i.e., each element of Z was multiplicatively distributed to each element of Z^{-1}. (Although no addition was used in that process.)

is $k \cdot A$, is not important; it has the form of an integer, and therefore, it is an integer.

Two other cases still need to be examined: the case where both factors are negative integers, and the case where one (or both) of the factors is zero. However, the results in either of these cases will also be an integer, although the latter case is trivial, because zero times any number is zero—an integer.

By extension, we will have also shown that the product of any two multiplicative inverses of integers is a multiplicative inverse of an integer. (I will leave all the details up to you, which could involve Theorem 3.5.3.)

◆ THEOREM: PRODUCT OF INTEGERS

The product of two integers is an integer, and the product of two multiplicative inverses of integers (unit fractions) is a multiplicative inverse of an integer (a unit fraction).

We can now show that the set $Z \otimes Z^{-1}$ is closed under multiplication. First, because $1 \in Z$ and $1 \in Z^{-1}$ (recall that the number 1 is its own multiplicative inverse), it follows that $Z \subset Z \otimes Z^{-1}$ and $Z^{-1} \subset Z \otimes Z^{-1}$.*

Second, if $a, b \in Z \subset Z \otimes Z^{-1}$, and $(1/c)$, $(1/d) \in Z^{-1} \subset Z \otimes Z^{-1}$, then, by the definition of $Z \otimes Z^{-1}$, it follows that $[a \cdot (1/c)]$ and $[b \cdot (1/d)] \in Z \otimes Z^{-1}$, from which we derive the following conclusion:

$$[a \cdot (1/c)] \cdot [b \cdot (1/d)] = (ab) \cdot (1/cd) \in Z \otimes Z^{-1} \quad (c,\ d \neq 0).$$

The expression to the right of equals is the product of an integer and the multiplicative inverse of an integer because ab and cd are both integers by the Product of Integers theorem; thus, $Z \otimes Z^{-1}$ is closed under multiplication

Finally, we can show that the sum of any two numbers from the set $Z \otimes Z^{-1}$ is a number in the set $Z \otimes Z^{-1}$, but first we must consider the following theorem about numbers in general.

◆ THEOREM 4.6.3

For any numbers a, b, c and d, where $b \neq 0$ and $d \neq 0$, then $\dfrac{a}{b} + \dfrac{c}{d} = \dfrac{ad + bc}{bd}$

Proof: (the details—the words—of the proof are yours to supply)

$$\frac{a}{b} + \frac{c}{d} = \frac{a}{b} \cdot 1 + 1 \cdot \frac{c}{d} = \frac{a}{b} \cdot \left(d \cdot \frac{1}{d}\right) + \left(b \cdot \frac{1}{b}\right) \cdot \frac{c}{d} = \frac{a}{b} \cdot \frac{d}{d} + \frac{b}{b} \cdot \frac{c}{d} = \frac{ad}{bd} + \frac{bc}{bd} =$$

$$ad\frac{1}{bd} + bc\frac{1}{bd} = (ad + bc)\frac{1}{bd} = \frac{ad + bc}{bd}.$$

It follows immediately from the Product and Sum of Integers theorems that if a, b, c and d ($b \neq 0$ & $d \neq 0$) are all integers, then $ad + bc$ and bd are integers.

* because the number 1 pairs, multiplicatively, with every number in Z and every number in Z^{-1}. (Note that for $P_{13} \otimes P_{79}$, $P_{13} \not\subset P_{13} \otimes P_{79}$ and $P_{79} \not\subset P_{13} \otimes P_{79}$. Why? (*Hint: the number* 1)

Consequently, $\frac{a}{b}$ and $\frac{c}{d} \in Z \otimes Z^{-1}$ implies that $\frac{a}{b} + \frac{c}{d} = \frac{ad + bc}{bd} \in Z \otimes Z^{-1}$. Thus, the set $Z \otimes Z^{-1}$ is closed under addition, and from an earlier discussion, we may now officially claim that $Z \otimes Z^{-1} = Q$. We may also write our discovery system as $[Q,+,\bullet]_A$ (i.e., $[Z \otimes Z^{-1},+,\bullet]_A$ and $[Q,+,\bullet]_A$ are identical). With that, we have reached the end of the line. There are no more authentic numbers that can be extracted from the axioms. This means, we may only play the game of algebra with either the set of rational numbers or with any "contrived" sets of objects that can be shown to adhere to all of the axioms.

However, mathematicians had many good reasons to extend the set of "playable" numbers to include the set of irrational numbers and the set of imaginary numbers as well. We will discuss the irrational numbers and a few other aspects of modern algebra in the next chapter.

We have only one thing left to do in this chapter, we must order the set of rational numbers. However, before we do that, a little perspective on the historical significance of the distributive property is in order.

It is often thought that mathematicians, unlike the theoreticians of the physical sciences, have no empirical data from which to derive their mathematical theories. Nothing could be further from the truth. A case in point is the historical derivation of a positive number from the product of two negative numbers.

Long before the arrival of the modern axiomatic method that we have today, mathematicians had suspected that the product of two negative numbers would have to be a positive number. Their suspicion was based on the recognition that the distributive property was undeniably true (Figure 22) for positive integers and if mathematicians were going to admit negative numbers into

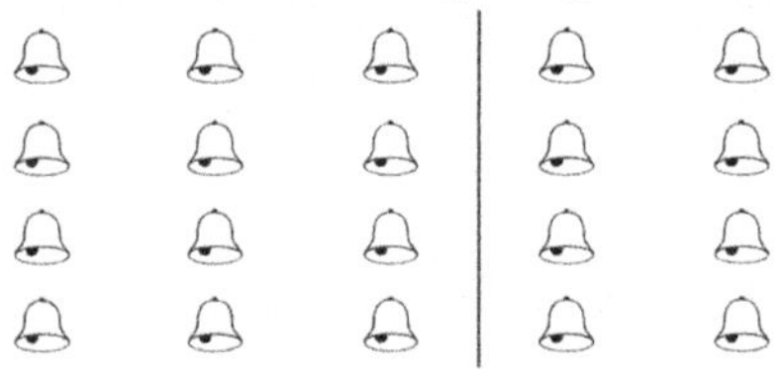

Figure 22: The distributive property: four rows of five columns of bells is the same as four rows of three columns of bells and four rows of two columns of bells.

the realm of real, existent numbers, then one concession or another would have to be made.

Consider, if you will, Figure 23 on the opposing page. A 5×7 array of bells is shown divvied up into four sections as indicated. There are several methods by which one can calculate the number of bells in any given sector, however, we will be concerned with just two methods for calculating the number of bells in the first sector (i.e., sector I).

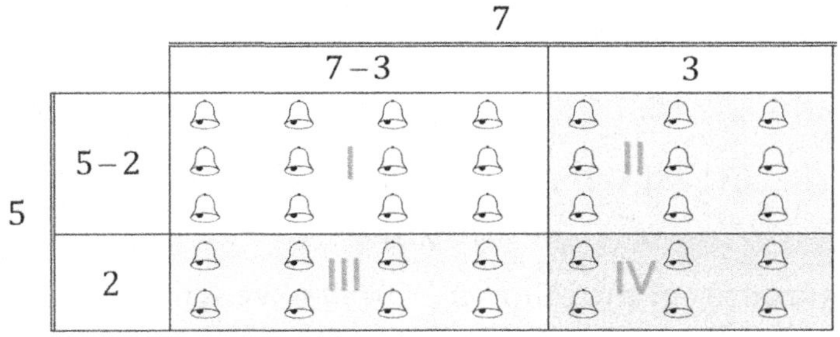

Figure 23: The distributive property and negative numbers.

Because subtraction was defined, in the past as it is now*, as the addition of negative numbers, we may write $5-2$ as $5+(-2)$ and $7-3$ as $7+(-3)$. Thus, if we wanted to find the number of bells in sector I, we would simply multiply the number of rows of bells by the number of columns of bells, which in this case for sector I is $(5-2)\times(7-3)=(5+(-2))\times(7+(-3))$. Using the distributive property to determine this product, we would arrive at the following four terms:

$$5\times7+(-2)\times7+5\times(-3)+(-2)\times(-3).$$ ▪Method 1

Alternatively, we could first calculate the total number of bells in all sectors, which is 5×7 and subtract from it all the bells which are in the union of sectors III and IV, and in the union of sectors II and IV, the result of which is $5\times7-2\times7-5\times3$. However, as you may have noticed, a mistake was made, we subtracted the bells from sector IV twice, or rather we subtracted 2×3 twice because that product appears in each of the two unions; so, we must add it once back. The result, using this method to calculate the number of bells in sector I, is the four terms,

$$5\times7-(2\times7)-(5\times3)+(2\times3).$$ ▪Method 2

Making a term by term comparison between the two methods we have,

	Term 1	Term 2	Term 3	Term 4
Method 1	5×7	$(-2)\times7$	$5\times(-3)$	$(-2)\times(-3)$
Method 2	5×7	$-(2\times7)$	$-(5\times3)$	2×3

If we want each method to yield the correct result for the number of bells in sector I, then we must have $(-2)\times7=-(2\times7)$, $5\times(-3)=-(5\times3)$ and, ultimately, we must have $(-2)\times(-3)=(2\times3)$. Hence, a negative number times a positive number *must* be a negative number, and a negative number times a negative number *must* be a positive number. Otherwise, we must ignore negative numbers altogether, deny the distributive property for some numbers, or let algebra and arithmetic become infested with contradictions. The choice was clear.

For the remainder of this book it will be helpful to list the sets we have used throughout its course (the order shown in each set may not reflect the actual order of the numbers):

* It still has some variations among mathematicians today.

$$Z_P = \{ 1, 2, 3, 4, \dots \};$$
$$Z_N = \{-1, -2, -3, -4, \dots \};$$
$$Z = Z_P \cup Z_N \cup \{0\};$$
$$Z^{-1} = \{(-1)^{-1}, (-2)^{-1}, (-3)^{-1} \dots 3^{-1}, 2^{-1}, 1^{-1}\}, \text{ and}$$
$$Q = Z \otimes Z^{-1}.$$

And, for the purposes of the following proofs, we will consider two other sets,

$$Z_P^{-1} = \{1^{-1}, 2^{-1}, 3^{-1}, 4^{-1}, \dots \}$$
$$Z_N^{-1} = \{(-1)^{-1}, (-2)^{-1}, (-3)^{-1}, (-4)^{-1}, \dots \}.$$

These last two sets are just the subsets of the positive numbers and the negative numbers of the set Z^{-1} respectively.

However, it is not necessary to read the remainder of this section regarding the logical order of the rational numbers; it is sufficient that you already intuitively know the order and that an order can, if desired, be logically imposed upon the rational numbers. It is provided here only because any, even moderately rigorous mathematical construction of the rational numbers requires it and it would be irresponsible of me not to include it. So, please feel free to skip the remainder of this section.

To deduce the order of the rational numbers, we will need Axiom O3 (the addition property for inequalities), but before that axiom will be of any use, we will first need the following helpful theorem:

◆ THEOREM 4.6.4

If $0<a$, *then* $0<\frac{1}{a}$; *and if* $a<0$, *then* $\frac{1}{a}<0$ *

Proof: (I will prove the first part by contradiction. The second part is up to you.) Suppose that $0<a$ and $1/a<0$, then because $0<a$ (and therefore $a \neq 0$), we have $a \cdot (1/a) < a \cdot 0$ (multiplication property of inequalities, E5), which leads to the (already proven) false statement $1<0$. Therefore, $(1/a) \not< 0$, and $1/a \neq 0$ (otherwise, $1 = a \cdot (1/a) = 0$, contrary to the fact that $1 \neq 0$); consequently, $0 < 1/a$ by the law of trichotomy, which is the required result.

To show order among the rational numbers we will consider two cases. Case 1 will show that if two integers are positive and one is less than the other, then their multiplicative inverses are positive and less than (or equal to) 1, and that the multiplicative inverse of the greater integer is less than the multiplicative inverse of the lesser integer (which is more easily understood in the language of mathematics). Case 2 will show an analog to Case 1 using negative integers.

---

* Although this theorem appears obvious, it does so only because of your training. For the purposes of logic, however, we must prove it.

Case 1: If $a,\ b \in Z_P$ and $a < b$, then $0 < \frac{1}{b} < \frac{1}{a} \le 1$.

Proof: Because $a,\ b \in Z_P$, we have $0 < a$ and $0 < b$, and therefore (by Theorem 4.6.4) $0 < \frac{1}{a}$ and $0 < \frac{1}{b}$. Because $a < b$, we may write $\frac{1}{a} \cdot a < \frac{1}{a} \cdot b$ which is equivalent to $1 < \frac{1}{a} \cdot b$ (multiplicative property of inequalities and axioms M5 and M4); also, $1 \cdot \frac{1}{b} < \frac{1}{a} \cdot b \cdot \frac{1}{b}$ is equivalent to $\frac{1}{b} < \frac{1}{a}$. However, zero is less than both of these numbers, and because 1 is less than or equal to every number $x \in Z_P$ (in symbols $1 \le x \in Z_P$) this case implies that $\frac{1}{x} \le \frac{1}{1} = 1$. Hence for every $x \in Z_P$, $\frac{1}{x} \le 1$. Therefore $0 < \frac{1}{b} < \frac{1}{a} \le 1.$*

Thusly, all the multiplicative inverses of the positive integers are ordered, from lesser to greater (reading from left to right) as indicated below. They are all greater than 0, but less than or equal to 1. That is,

$$0 < \ \cdots\ \frac{1}{4} < \frac{1}{3} < \frac{1}{2} < \frac{1}{1} = 1\,.$$

Case 2: If $a,\ b \in Z_N$ and $a < b$, then $-1 \le \frac{1}{b} < \frac{1}{a} < 0$.

Proof: Because $a,\ b \in Z_N$, we have $a < 0$ and $b < 0$ and therefore (by Theorem 4.6.4) $\frac{1}{a} < 0$ and $\frac{1}{b} < 0$, which implies $0 < -\frac{1}{a}$ and $0 < -\frac{1}{b}$ (Why?). Then, because $a < b$, we have $\left(-\frac{1}{a}\right)a < \left(-\frac{1}{a}\right)b$, which is just $(-1) < \left(-\frac{1}{a}\right)b$ (Why?). This last statement, along with the fact that $0 < -\frac{1}{b}$, allows us to write $(-1)\left(-\frac{1}{b}\right) < \left(-\frac{1}{a}\right)b\left(-\frac{1}{b}\right)$, which implies $(-1)\left(-\frac{1}{b}\right) < \left(-\frac{1}{a}\right)(-1)$, but a negative times a negative is a positive; hence $\frac{1}{b} < \frac{1}{a} < 0$. Because $x \le -1$ for all $x \in Z_N$, $-1 = \left(\frac{1}{-1}\right) \le \frac{1}{x}$ for all $x \in Z_N$; therefore $-1 \le \frac{1}{b} < \frac{1}{a} < 0$.

Thusly, all the multiplicative inverses of the negative integers are ordered from lesser to greater (reading from left to right) as indicated below. They are all greater than or equal to -1, but less than 0.

$$-1 = -\frac{1}{1} < -\frac{1}{2} < -\frac{1}{3} < -\frac{1}{4},\ \cdots\ < 0$$

---

* In higher math, this involves the concepts of least upper and greatest lower bounds.

We can now write, affirmatively, the order of the mix of the multiplicative inverses of the integers (except zero) together with the integers from lesser to greater:

$$\ldots -4,\ -3,\ -2,\ -1,\ -\frac{1}{2},\ -\frac{1}{3},\ -\frac{1}{4},\ \cdots\ 0,\ \cdots\ \frac{1}{4},\ \frac{1}{3},\ \frac{1}{2},\ 1,\ 2,\ 3,\ 4\ldots$$

Of course, there are many (infinitely many) more fractions between each successive integer, numbers like $-21/16$, $13/5$ and so on. And there are many (infinitely many) more fractions between -1 and 1 that we haven't listed, numbers like $2/3$, $-79/101$ and so on. Fortunately, ordering these numbers within the ordered set that we currently have is a relatively straight forward procedure. (I will just outline it here by way of example, you will need to supply the more general details, if you care to, but faith, for our purposes is sufficient.)

It is easy to show for any number from the set Z_P^{-1}, say $1/23$, and any distinct pair of successive numbers from the set Z_P, say 87 and 88, that $87 < 87 + 1/23 < 88$. To begin, $0 < 1/23 < 1$ (which we have shown to be true) implies that $87 < 87 + 1/23$ and $87 + 1/23 < 87 + 1$, but by definition $87 + 1 = 88$; hence, by substitution, $87 + 1/23 < 88$ and by the transitive property of inequalities we have $87 < 87 + 1/23 < 88$. The equivalent number for $87 + 1/23$ is found by Theorem 4.6.3, which happens to be the improper fraction $2002/23$ or by convention we would write this as the proper fraction $87\,{}^{1}\!/_{23}$.

Also, if we go back to our original definitions and ordering of the integers,

$$\cdots(-1)+(-1)+(-1)<(-1)+(-1)<(-1)<0<1<1+1<1+1+1\cdots,$$

then, if we multiply "through" by some number from the set Z_P^{-1}, say $1/7$, then, with the help of the distributive axiom (among others), we can show that

$$\cdots\ -\frac{3}{7}<-\frac{2}{7}<-\frac{1}{7}<0<\frac{1}{7}<\frac{2}{7}<\frac{3}{7}\ \cdots$$

Of course, we could do this with any number from the set Z_P^{-1}.

It is important to keep in mind, however, that the axioms of algebra were written to match, or at least not to conflict with, our understanding of elementary arithmetic, and every axiom was needed throughout the course of this book to bring that about. Remove any one of them and we would not be able to deduce elementary arithmetic, never mind the fact that the great utility of algebra would come up woefully short.

~

CHAPTER 4 EXERCISES

1. Determine which of the following are closed sets under multiplication. If a set is not closed, produce a counter example.
   a. $\{1\}$ b. $\{-1,0\}$ c. $\{1,0\}$ d. The set of odd numbers.
   e. The set of all elements multiplicatively generated from the elements 0, 2 and -2. Describe the set generated. Is the set a group under multiplication?
2. Show that the set of sets $S=\{\varnothing, \{a\}, \{a,b\}, \{a,b,c\}\}$ is (a) a closed set under the operation of intersection and union; (b) explain why $\varnothing \in S$ and $\varnothing \subset S$; (c) is the set a group under either operation?
3. Using the 12 numbers on a clock and wrap-around multiplication, generate the set from the given set of elements and determine (1) if the generated set is a group under multiplication, (2) if the generated set is a subset of C (the set of elements on a clock), and (3) which, if any, of the generated sets are equal.
   a. $\{1,2\}$ b. $\{1,3\}$ c. $\{1,5\}$ d. $\{1,6\}$
   e. $\{1,7\}$ f. $\{1,9\}$ g. $\{1,10\}$ h. $\{1,11\}$
4. What set is generated from the number 7 using wrap-around multiplication with the number 13? (Thus, $7\cdot 7=10$, etc.)
5. Prove that $2+3=5$ and that $2-3=-1$.
6. Show or explain why
   a. $\dfrac{y-2}{2-y}=-1$ b. $(a/b)/(c/d)=(ad)/(bc)$ e. $1\cdot 0=0$
7. Using just two instances of Theorem 4.6.3 and no other axioms, show that
$$\frac{a}{b}+\frac{c}{d}+\frac{e}{f}=\frac{(ad+bc)f+(bd)e}{(bd)f}$$
8. Simplify and be mindful of the axioms in play.
   a. $\dfrac{2}{3}+\dfrac{7}{6}+\dfrac{11}{12}+\dfrac{13}{24}$ b. $\dfrac{3t+7t+5}{4t+16t+10}$ c. $\dfrac{6y-3}{4y-2}+\dfrac{3z-9}{33-11z}$
   d. $\dfrac{3a-2b}{2b}-\dfrac{3x-2y+7t}{7t}$ e. $\dfrac{\frac{-ay}{x}+4y}{y}-\left(\dfrac{a}{-x}-3x\right)$
   f. $\dfrac{2n-k}{h}\cdot\dfrac{(2x+1)h}{x}\cdot\dfrac{z+1}{6n-3k}$ g. $4(7h^{-1})^{-1}-\dfrac{h}{7}$ h. $(x+\pi)(x-\pi)$
   i. $(2x+3y+4)^{-1}(2x+3y-4)^{-1}$ j. $(y-2x)^{-1}/[-7\cdot(14x-7y)^{-1}]$
9. The axioms only imply the existence of rational numbers, but what about irrational numbers, like $\sqrt{2}$ or π, how are we to account for these numbers? What does your intuition tell you about there being other real numbers besides just the rational and irrational numbers? Could there be something between the two?

CHAPTER 4 EXERCISES

1. Determine which of the following are closed sets under multiplication. If a set is not closed, produce a counter example.

[illegible]

CHAPTER 5

IRRATIONAL AND REAL NUMBERS, ROOTS AND EXPONENTS

5.1 ROOTS, IRRATIONALS, AND REAL NUMBERS

"Please stop inviting the Irrationals, all they ever do is argue, and I wish the Imaginaries would stop eating all the Marciano cherries."
"But dear, I didn't invite the Imaginaries."

We all know that algebra involves more than just the rational numbers. It also involves the irrational numbers, like $\sqrt{2}$ and π, but these numbers cannot be deduced directly from the axioms. However, the real utility and power of algebra is not in what can be directly deduced from its axioms, if that were the case, we would be left with nothing more than elementary arithmetic, rather it is in what we can discover through assumptions and interestingly crafted equations. For example, if a is a number, then we know that $a \cdot a$ is a number, that is axiomatic and nothing new; however, if we assume that $a \cdot a = 9$ (because we simply cannot deduce this equation) then we are faced with a small problem, which asks, "what number times itself is 9?"

Most of you, I am sure, are saying, "that's easy, the number is 3 or -3 because $3 \cdot 3 = 9$ and $(-3) \cdot (-3) = 9$". You would be right, of course, except we still would not know, *logically*, whether a from the equation $a \cdot a = 9$, in fact, equals 3 or -3. However, we can show that $a \cdot a = 3 \cdot 3$ using the transitive property of equality; then, by virtue of the first theorem of this book (Theorem 2.1.1), we can conclude that $a = 3$, (or $a = -3$). Likewise, if $a \cdot a \cdot a = 125$, then, because $125 = 5 \cdot 5 \cdot 5$, we can conclude that $a = 5$, by Theorem 2.1.1 Extended.

On the other hand, if we assume that $a \cdot a = 2$, then determining the value of a in this case has some very interesting consequences. Unfortunately, we can't use Theorem 2.1.1 to determine the actual value because that theorem depends on first knowing the number that when multiplied by itself equals 2.

Since we have already developed the rational numbers from the axioms, we may consider taking some rational number and see if when it is multiplied by itself it equals 2; and knowing that 2 is too large ($2\times 2=4$) and 1 is too small, we may try a number between the two, say $3/2$, for starters. However, the product of $3/2$ multiplied by itself is $9/4$, which is slightly larger than 2. So we try a number slightly smaller than $3/2$, perhaps $(2\tfrac{1}{2})/2=5/4$. However, $5/4$ times itself is $25/16$, which is slightly less than 2. Hence, we shrewdly try a number slightly larger than the last, but smaller than the one before it, say $(2\tfrac{3}{4})/2$ (splitting the difference is always a good strategy). We could continue in this fashion indefinitely, but eventually we would realize that we are not getting anywhere and begin to wonder whether such a number exists. In fact, it turns out, there is no rational number that when multiplied by itself is *exactly equal* to 2 and there is a proof, worn-out as it is, of it.

The proof (the architect of which is lost to history) involves, once again, that marvelous hammer of logic, *proof by contradiction**. However, before we look at a proof proper, we need to recall the definition of even and odd numbers from a footnote of an earlier section (good definitions are essential to the axiomatic process) and we will also need the following two lemmas. (Lemmas are precursor theorems to more important theorems.)

Lemma-1: Even times even is even. Previously, we relied only on our good mathematical intuition for the conviction that the product of any two even numbers is an even number. Now, however, we will prove it using the following (agreed upon†) definition: an even number has the form $2x$, such that $x\in Z$. Thus, if $2x$ and $2y$ are even numbers (and hence, integers), then their product is an integer (previously proved) and $(2x)(2y)=2\cdot(2xy)$, which happens to have the form of an even number; that is to say, it is twice some integer (the integer that is being doubled here is $2xy$). Therefore, the product of even numbers is an even number.

Lemma-2: Odd times odd is odd. An odd number, by definition, has the form $2x+1$, such that $x\in Z$. The set of odd numbers is closed under multiplication because if $2x+1$ and $2y+1$ are odd numbers (and hence, odd integers); then, with the help of the distributive property, we have,

$$(2x+1)(2y+1)=(2x)(2y)+2x+2y+1=2(2xy)+2x+2y+1=2\underbrace{(2xy+x+y)}_{\text{This is an integer}}+1.$$

The last expression has the form of an odd number (i.e., it is twice an integer plus one). Therefore, the product of odd numbers is an odd number.

Having gotten the preliminaries out of the way, we may now proceed with

* However, direct proofs are always preferred whenever it is possible to do so.

† Other definitions are used, e.g., any number which leaves no remainder when divided by 2.

the proof proper. Because this is a proof by contradiction, we assume the opposite conclusion, which is that there *is* a rational number a, such that $a \cdot a = 2$. We will write this rational number as b/c; thus, $(b/c)\cdot(b/c)=2$ (b and $c>0$). We will also agree that b/c is in reduced form.* Thus, after performing the multiplication, we have $(bb)/(cc)=2$ and upon multiplying both sides of this equality by (cc), we obtain $bb=2cc$. Consequently, bb is an even number†. However, this implies that the number b is even because if it were odd, the product of b with itself would be odd. Consequently, b must be equal to two times some integer, say $b=2k$ (where $k \in Z_P$). After substituting what b is equal to in the equation $bb=2cc$, we have $(2k)(2k)=2cc$. However, $(2k)(2k)=2(2kk)$. Substituting again, we have $2(2kk)=2cc$, and applying the multiplicative inverse of 2 to both sides, we have $2kk=cc$, which, by the same reasoning, means that c is even. However, we agreed that b and c had no common factors, yet we have just shown that they both have the number 2 as a common factor. This is a contradiction. Hence, our assumption that there is some rational number that when multiplied by itself equals 2 is false.

The axioms imply that if there is no *rational* number a that will satisfy the equation $a \cdot a = 2$, then no such number exists because rational numbers are the only numbers producible from the axioms. Recall that Axiom M1 stated that *if a is a number, then $a \cdot a$ is a number*; in other words, unless a is a number there is no guarantee that $a \cdot a$ is a number. The alpha-number a was always just a pseudo-number—a stand in or logical variable for the real thing. Furthermore, the converse of Axiom M1—i.e., *if ab is a number, then a and b are numbers*—was never stipulated as an axiom and cannot be proved based on the axioms alone.‡ Therefore, because a is not a rational number and $a \cdot a = 2$ (although, itself, an integer number) is insufficient to imply that a is a number, we have no logical means of determining whether or not a is a number. (Nor have we introduced a "tidy" symbol (see Def. 5.1.2 for one) for what a means in the context of $a \cdot a = 2$, we have been using the English language to describe it.)

Consequently, we must consider one of two possibilities: (1) forcibly include those numbers that are not rational, otherwise known as irrational numbers, in with the collection of numbers we have already acquired through the axioms; or (2) expel irrational numbers from our thinking altogether, and never

---

* b and c ($\neq 0$) are, for our purposes, positive integers, and we can agree that b and c have no common factors because whatever fraction may have equaled b/c all of its common factors have been canceled out, leaving us with just b/c —the lowest-term fraction. Thus, in particular, b and c cannot both be even numbers.

† because it is equal to 2 times an integer, which is an even number.

‡ The converse was not stipulated because if it were, then when we multiplicatively close the set $\{1\}$, or any set for that matter, it would always equal the set of real numbers because the equality axioms would allow entry of all the products of $1=(1/2)\cdot 2=(1/3)\cdot 3$, plus everything else in between and beyond, totally wrecking more than a century of work on the theory of sets.

be bothered with them again.

However, much to their great disappointment, mathematicians had no choice but to include them because, not surprisingly, there exist many well-known numbers which are irrational. The diagonal of a square, one unit on edge (i.e., $\sqrt{2}$), is a prime example of a number that is not rational (a fact that gnawed at the Ancient Greeks' cherished belief in the harmony of numbers right up until the demise of their civilization). The Golden Ratio is another example. In Figure 24, if $(a+b)/a = a/b$, then $a/b = 1.618...$ —the Golden Ratio.

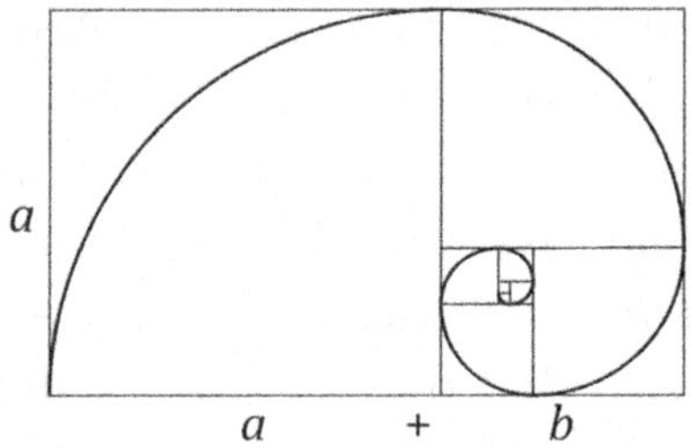

Figure 24: The Golden ratio: $a/b = 1.618...$

The ratio of the circumference of a circle to its diameter, π, is yet another. However, a formal definition of an irrational number is long overdue, so please allow me a moment to introduce one.

∘ DEF. 5.1.1: IRRATIONAL NUMBERS

An irrational number is any number that <u>*cannot*</u> *be expressed as the quotient of two integers.*

However, this definition, which is commonly found in high school texts, lacks some critical elements. It doesn't *directly* define irrational numbers, how to find them, how they are ordered, or even if they exist—although we have just proved they do.* It only tells us that if we have a "known" number that is not rational, then it is irrational. (Hence, pure imaginary or complex numbers with imaginary parts [see section 6.5] are irrational. [Why exactly?])

In the minds of mathematicians, however, this is an unacceptable and somewhat empty definition. What, after all, does *any number* in the definition mean, and how do we know when we have a number? Is every number-like thing we can put to paper, however bizarre, a number? Is $\sqrt[-\pi]{-e^{i\pi}}$ a number, is it rational, irrational, imaginary or something else?† Apropos, there are numbers which have not (to date) been proven as either rational or irrational, $\pi^{\sqrt{2}}$ for example. There are also numbers which balance on the precipice of being either.

Out of necessity, therefore, mathematicians had to devise methods to construct, order and define the set of irrationals. Although the methods of Richard Dedekind, called Dedekind Cuts, and Cauchy sequences are probably the two most popular methods for this construction and not exceptionally difficult to

---

* Without having or proving the existence of at least one irrational, the definition is pointless.

† Just because a number involves an imaginary does not mean the number itself is imaginary.

understand, they are beyond the scope of this book, so we will make do with the more traditional, elementary definition supplied above.

However, we are not without recourse, we have at least one irrational number to work with: *the number that when multiplied by itself exactly equals* 2. However, we have no good way to exhibit this number *exactly*, we don't know how to write or communicate it beyond this tedious and clumsy use of English—something simpler is demanded—hence the following definition:

∘ DEF. 5.1.2: THE SQUARE ROOT/RADICAL OF A NUMBER

The square roots of a number, $a \geq 0$, *written as* $\pm\sqrt{a}$ *or* $\pm\sqrt[2]{a}$, *are the numbers that when multiplied by themselves exactly equal the number* a. *Thus, because* $\sqrt{a}\cdot\sqrt{a}=a$ *and* $(-\sqrt{a})\cdot(-\sqrt{a})=a$, $\sqrt{a}$ *and* $-\sqrt{a}$ *are square roots of* a. $\sqrt{a}$ *is called the principal square root of* a *– it is always a positive number.*

Consequently, we may now write $\sqrt{2}\cdot\sqrt{2}=2$ and, as we have already shown, $\sqrt{2}$ is irrational. Of course, any other symbol could have been used to convey the meaning of square root, but "$\sqrt{\ }$", known as the *radical symbol*, is the symbol passed down through history. The number under the radical is called the *radicand*, while the number wedged in the upper left-hand corner of the radical symbol is called the *index*. The index is always a positive integer, making radicals somewhat less convenient to work with than exponents.

Of course, we do not need to limit ourselves to just the square root of 2, we can examine the square root(s) of any number, thus we may write $\sqrt{5}\cdot\sqrt{5}=5$, $\sqrt{9}\cdot\sqrt{9}=9$, and $\sqrt{ab}\cdot\sqrt{ab}=ab$.* However, on careful examination of the above definition, we would have to agree, from an earlier discussion, that 3 is also a square root of 9 because 3 is a number that when multiplied by itself equals 9. However, because $\sqrt{9}\cdot\sqrt{9}=9$ and $3\cdot 3=9$, we have $\sqrt{9}\cdot\sqrt{9}=3\cdot 3$, which (Theorem 2.2.1) implies $\sqrt{9}=3$; and because $0\cdot 0=0$ and $\sqrt{0}\cdot\sqrt{0}=0$, we also have $\sqrt{0}=0$ as a *logical* consequence of this definition.

However, there is a slight glitch in the language when using the definition of square roots (and for all even roots) that many teachers do not satisfactorily address. This, unfortunately, has led to the confusion of countless students whenever the subject of roots is raised. The reason for the confusion is that the number -3 is also a "square root" of 9, as we have been using the language. However, it would be incorrect to write $\sqrt{9}=-3$. Doing so would be like writing $3=-3$. The correct language to use is $\sqrt{9}=3$, $-\sqrt{9}=-3$; hence, the inclusion of the caveat of a *principal square root* in the definition. The only other way to resolve this glitch is to use English and write "the *square roots* of 9 are 3 and

---

* It is interesting to note that even as late as the middle of the 19th century mathematicians had not yet proved something as "simple" as $\sqrt{3}\times\sqrt{5}=\sqrt{3\times 5}$, even though the empirical evidence for such equations was overwhelming. The difficulty with the proof was not for lack of intuition or understanding, but rather for lack of a strong axiomatic structure with good definitions from which a proof could be deduced.

-3," (meaning, those numbers that when multiplied by themselves equal 9), but mathematicians are much too lazy to be bothered with all that.

This is not a problem for roots with odd indices because there is only one (real) number that can be equal to such a root. In other words, there is only one cube (3rd) root of a number; thus, $\sqrt[3]{27}=3$, and this is the only number that when multiplied by itself three times is 27. Thus, in contrast to roots with even indices, we may write $\sqrt[3]{-27}=-3$ and $-\sqrt[3]{27}=-3$ when the index is odd. Hence, by the transitive property, we have $\sqrt[3]{-27}=-\sqrt[3]{27}$. In general, when the index n is odd $\sqrt[n]{-a}=-\sqrt[n]{a}$ (more about roots with indices greater than 3 shortly).

Also, because $\sqrt{a}$ is a number (provided that $a\geq 0$ is a number), it has an additive inverse $-\sqrt{a}$ (which we have already discussed) and a multiplicative inverse $1/\sqrt{a}$ (when $a\neq 0$).

Having made the definition, however, we need to understand how we can use it to simplify our lives when doing algebra—we need to determine how two or more different square roots add and multiply. The next theorem states that *the product of the square roots of two numbers is equal to the square root of the product of the two numbers.*

◆ **THEOREM 5.1.1**

If $0\leq a$ *and* $0\leq b$ *then* $\sqrt{a}\cdot\sqrt{b}=\sqrt{ab}$

Proof: $(\sqrt{a}\cdot\sqrt{a})\cdot(\sqrt{b}\cdot\sqrt{b})=ab$ and $(\sqrt{a}\cdot\sqrt{a})\cdot(\sqrt{b}\cdot\sqrt{b})=(\sqrt{a}\cdot\sqrt{b})\cdot(\sqrt{a}\cdot\sqrt{b})$, so $(\sqrt{a}\cdot\sqrt{b})\cdot(\sqrt{a}\cdot\sqrt{b})=ab$. It is also true (by definition) that $\sqrt{ab}\cdot\sqrt{ab}=ab$, therefore $(\sqrt{a}\cdot\sqrt{b})\cdot(\sqrt{a}\cdot\sqrt{b})=\sqrt{ab}\cdot\sqrt{ab}$. Consequently, by Theorem 2.1.1, we can conclude that $\sqrt{a}\cdot\sqrt{b}=\sqrt{ab}$. (The case where either number equals 0 is trivial.) As a practical example of this theorem, we can determine that $\sqrt{2}\cdot\sqrt{18}=\sqrt{2\cdot 18}=\sqrt{36}=\sqrt{6\cdot 6}=\sqrt{6}\cdot\sqrt{6}=6$.

We may also wonder what number a would make the statement $aaa=2$ true; that is, what number multiplied by itself three times is the number 2? Or for that matter, what number a would make the statement $aaaa=2$ true? Again, in most cases we cannot find the exact number, so we must develop the language—the words, if you will—necessary to discuss such exact numbers.

∘ DEF. 5.1.3: THE nth ROOT/RADICAL OF A NUMBER

The nth *root of a number* a *(*$n\in Z_P$, $n\geq 2$, *and* $a\geq 0$ *if* n *is even), written as* $\sqrt[n]{a}$, *is the number that when multiplied by itself* n *times exactly equals* a. *Thus,*

$$\underbrace{\sqrt[n]{a}\cdot\sqrt[n]{a}\cdot\sqrt[n]{a}\cdot\ \cdots\ \sqrt[n]{a}}_{n\text{ factors of }\sqrt[n]{a}}=a.$$

Consequently, $\sqrt[3]{2}\cdot\sqrt[3]{2}\cdot\sqrt[3]{2}=2$ and $\sqrt[4]{79}\cdot\sqrt[4]{79}\cdot\sqrt[4]{79}\cdot\sqrt[4]{79}=79$, etc.

The next theorem follows the form of the proof for Theorem 5.1.1 but involves the extension to Theorem 2.2.1. However, we will not prove it. (I will leave it for you to express the mathematical symbols into English.)

◆ **THEOREM 5.1.2**

$\sqrt[n]{a}\cdot\sqrt[n]{b}=\sqrt[n]{ab}$ *($n\in Z_P$ and $n\geq 2$), provided $0\leq a$ and $0\leq b$ whenever n is even*

There is no corresponding general theorem for the addition of roots, in other words, $\sqrt[n]{a}+\sqrt[n]{b}\neq\sqrt[n]{a+b}$, which is relatively easy to prove. For now, however, we will only present the following counter example.

$$5=2+3=\sqrt[3]{8}+\sqrt[3]{27}\neq\sqrt[3]{8+27}=\sqrt[3]{35}\approx 3.27.$$

In order to abide by the closure axioms, we must include the roots of every number we have already "authenticated", as well as numbers like π and e*, which, although irrational, are unlike any of the numbers we have previously encountered (but beyond the scope of this book—called transcendental numbers, they "outnumber"† all the other numbers combined). All of the numbers in this expanded set must then be added and multiplied together in every conceivable fashion as well as the products and sums of those numbers, and the products and sums of those, and so on, to satisfy the demands of the closure axioms, after which, they must all be include in the set.

This monstrosity of a set is called the set of *real* numbers, which we denote with the letter R (often the symbol "ℝ" is used to represent the reals, as are the symbols "ℤ", "ℚ", and "ℕ" often used to represent the integers, the rational numbers and the natural numbers, respectively). The set R is best understood using the traditional number line (Figure 25).

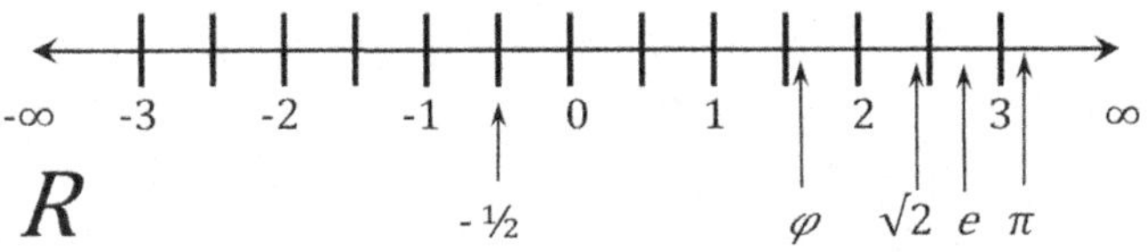

Figure 25: The real number line (the set of all real numbers).

The set of real numbers is considered a *continuum*—there is no demarcation where one number ends, and another begins. Nevertheless, mathematicians "pull out" and use numbers from the continuum as if they were singularities, as if they were points that existed in isolation from every other point. The square root of two (there is it is ➔ $\sqrt{2}$) is a perfect example.

Although we haven't supplied any *logical* order to these numbers, for our purposes, we will assume that they are ordered exactly as we have always known them to be. This is the position that every high-school algebra book takes when it comes to "defining" real numbers—it simply leaves it to your good intuition and the number line. However, I should reiterate that because this informal picture or mental image of the real numbers is only intuitive, it is unusable as a logical definition of the real numbers or as a logical definition of their order, hence the need for developing more rigorous definitions of the real

* The base of the natural logarithm is approximately equal to 2.71828.

† Some infinitely large sets are (theoretically anyway) larger than other infinitely large sets.

numbers such as those proposed by Dedekind and Cauchy.

But all that aside, and much to everyone's great disappointment, our once beautiful, tranquil, and logically understandable structure involving the axioms of algebra and the rational numbers has become a complete and utter mess. Numbers, it would seem, like the laws governing the universe, are not only mysterious, but also completely beyond our will and wishes.

~

5.2 EXPONENTS

By now I am sure you are all tired of writing products like *aaaa* and $\sqrt[5]{79}\cdot\sqrt[5]{79}\cdot\sqrt[5]{79}$, well, so were mathematicians; hence, they established the following abbreviation/definition:

∘ DEF. 5.2.1: EXPONENTS

The nth power ($n \in Z_P$ and $n \geq 2$) of a number a written as a^n, is the number a multiplied by itself n times. Thus,

$$\underbrace{a \cdot a \cdot a \cdot \ \cdots \ a}_{n \text{ factors of } a} = a^n.$$

The number n is called the *exponent* and the number a is called the *base* and we define $a^1 = a$.

Using this definition, it is easy enough to show that $a^k a^j = a^{k+j}$. Consequently, $3^5 3^3 = 3^{5+3} = 3^8$. In other words, we add the exponents when taking the product of numbers that have the same base. Furthermore, because Theorem 5.1.2 implies that $\sqrt[n]{a}\cdot\sqrt[n]{a} = \sqrt[n]{a \cdot a}$, it follows from Def. 5.2.1 that $(\sqrt[n]{a})^2 = \sqrt[n]{a^2}$. In general, we may write $(\sqrt[n]{a})^k = \sqrt[n]{a^k}$ (and in particular $(\sqrt[k]{a})^k = \sqrt[k]{a^k} = a$).

There is nothing stopping us from taking the power of a number which is already raised to a power, either. Thus, $(3^5)^3$ means to multiply 3^5 by itself 3 times, which is just $3^5 3^5 3^5 = 3^{15} = 14{,}348{,}907$. Thus, *raising to a power a number raised to a power is just the number raised to the product of the powers*. This, as usual, is easier said in the language of mathematics:

◆ **THEOREM 5.2.1**

$(a^m)^n = a^{m \cdot n}$ (☆10)

And if that's not complicated enough, then consider this, we can also raise a power to a power. However, it is important to be mindful of parentheses):

◆ **THEOREM 5.2.2**

$$a^{m^n} = a^{\overbrace{m \cdot m \cdot m \ \cdots \ m}^{n \text{ factors of } m}} \quad \text{(Note that } a^{m^n} \neq (a^m)^n.\text{)}$$

Thus, $3^{5^3} = 3^{5 \cdot 5 \cdot 5} = 3^{125}$, a 60-digit long number—a very big number, indeed.

When we introduced the concept of multiplicative inverses in Section 3.5, we considered the symbol a^{-1} as strictly representing the multiplicative inverse of a. Now, however, it would be convenient for us to think of the -1 in the upper right-hand corner as an exponent. Thus, because $a = a^1$ by definition, and $aa^{-1} = 1$ by axiom, it follows that $a^1 a^{-1} = 1$; furthermore, if we want to continue with the idea of adding exponents when taking the product of numbers with the same base, each of which has been raised to a power, then we must allow $a^1 a^{-1} = a^{1+(-1)} = a^0 = 1$ (except when $a = 0$, as we shall see). Therefore, every number (except 0) raised to the power of zero equals one.

Adding exponents, whether they are negative or positive, works well with the properties of multiplicative inverses. Thus, using Theorem 5.2.1, we have,

$$a^3 a^{-4} = a^3 a^{(-1)(4)} = a^3 (a^{-1})^4 = (aaa)\left(\frac{1}{a}\frac{1}{a}\frac{1}{a}\frac{1}{a}\right) = \left(a\left(a\left(a\frac{1}{a}\right)\frac{1}{a}\right)\frac{1}{a}\right)\frac{1}{a} = \frac{1}{a} = a^{-1} = a^{3-4}. \quad \blacksquare 5.2.1$$

Definition 5.2.1 can now be expanded to include negative exponents and the zero exponent, which makes working with and analyzing algebraic equations extremely easy. It is certainly easier as compared to writing out long strings of factors like those shown in line 5.2.1 above, and because exponents are numbers, they too abide by all the rules of algebra. (This repeating pattern of algebra carries down to the finest detail—it is what makes algebra easy.)

There is, however, one caveat about exponents: considering that $0^1 = 0$ is well defined, as is 0^2, 0^3, etc., and because 0^{-1} is *not* defined*, it should be clear that $0^1 \cdot 0^{-1} = 0^{1-1} = 0^0$ is also not defined, because we are using an undefined term in our calculation. (However, for reasons of convenience, some math formulas *define* $0^0 = 1$ or $0^0 = 0$, but we will leave it undefined.)

We can extend the idea of exponents even further if we allow exponents to take on rational values in addition to integer values, provided we think of the numerators as integer powers and the denominators as integer roots. With that in mind we have the following two definitions.

∘ DEF. 5.2.2: RADICALS AS EXPONENTS

$$a^{1/n} = \sqrt[n]{a}.$$

Theorem 5.1.1 then becomes $a^{1/2} \cdot b^{1/2} = (ab)^{1/2}$, and Theorem 5.1.2 becomes $a^{1/n} \cdot a^{1/n} = (ab)^{1/n}$.

∘ DEF. 5.2.3: DEFINITION OF EXPONENTS EXTENDED

The j/k power of a number a, written as $a^{j/k}$ (where $j, k \in Z$, $k \neq 0$ and $a \geq 0$ whenever k is even and j is odd†), is the kth root of the number a multiplied by itself j times. Thus, $a^{j/k} = (\sqrt[k]{a})^j$ or alternatively $a^{j/k} = \sqrt[k]{a^j}$.

Thus, we have two methods to consider when dealing with the roots of numbers: we can work with roots expressed as radicals (i.e., with the symbol $\sqrt{\ }$), or we can work with roots expressed as the denominators of exponents.

---

* Zero raised to any exponent less than zero is also undefined.

† This requirement is needed only to keep us out of the realm of imaginary numbers.

As a comparison, we will first consider the method of determining the product of $\sqrt[2]{a}\cdot\sqrt[3]{b}$ by using radicals and then consider the method of determining the product by using exponents.

First method: we observe that $(\sqrt[2]{a}\cdot\sqrt[2]{a})\cdot(\sqrt[3]{b}\cdot\sqrt[3]{b}\cdot\sqrt[3]{b})=(\sqrt[2]{a})^2\cdot(\sqrt[3]{b})^3=ab$. However, because we have three factors of $\sqrt[3]{b}$ and only two factors of $\sqrt[2]{a}$, it is impossible to create identical factors—a necessary condition if we wish to use Theorem 2.1.2. Observing further that $a^3\cdot b^2=(\sqrt[2]{a})^2(\sqrt[2]{a})^2(\sqrt[2]{a})^2\cdot(\sqrt[3]{b})^3(\sqrt[3]{b})^3$ allows us to create—six—identical factors of $\sqrt[2]{a}\cdot\sqrt[3]{b}$ (decide for yourself that this is indeed the case), we may rewrite the right-hand side of this equality as $(\sqrt[2]{a})^6(\sqrt[3]{b})^6=((\sqrt[2]{a})(\sqrt[3]{b}))^6=a^3b^2$. However, because $a^3b^2=(\sqrt[6]{a^3b^2})^6$, it follows from the transitive property of equality that $((\sqrt[2]{a})(\sqrt[3]{b}))^6=(\sqrt[6]{a^3b^2})^6$; thus, by Theorem 2.1.2, we have the required result: $\sqrt[2]{a}\cdot\sqrt[3]{b}=\sqrt[6]{a^3b^2}$.

Alternatively, using the second method, the product $\sqrt[2]{a}\cdot\sqrt[3]{b}$ could have been solved by first transforming the radicals into fractional exponents, where the numerator is an integer power and the denominator is an integer root; after which it is just a matter of transforming the fractional exponents into fractions which have a common denominator. Specifically, $\sqrt[2]{a}\cdot\sqrt[3]{b}=a^{1/2}\cdot b^{1/3}=a^{3/6}\cdot b^{2/6}=a^{3(1/6)}\cdot b^{2(1/6)}=(a^3)^{1/6}\cdot(b^2)^{1/6}$; this last expression is equal to $(a^3\cdot b^2)^{1/6}=\sqrt[6]{a^3b^2}$ by Theorem 5.1.2. Observe that this product was determined by finding equivalent fractions to $1/2$ and $1/3$, such that they have a least-common denominator; that is to say $1/2=3/6$ and $1/3=2/6$. From there, we simply used the rules of exponents. I think you would agree that fractional exponents are easier to work with than radicals.

As a result of the foregoing, we have verified the following theorem:

◆ THEOREM 5.2.3

$\sqrt[m]{a}\cdot\sqrt[n]{b}=\sqrt[(mn)]{a^nb^m}$ *or equivalently* $a^{1/m}b^{1/n}=(a^nb^m)^{1/(mn)}$. (☆11)

Of course, there is never a need to use radical expressions at all, we can perform and exhibit all of our calculations with just the concept of rational exponents. Using exponents also has the added benefit of being able to manipulate numbers with irrational exponents, as we shall see in Chapter 6. However, using the radical symbol, where appropriate, often makes a mathematical expression considerably more readable. When to use radicals, and when to use rational exponents is a matter of taste and experience.

Finally, at the risk of repeating myself, it is important you understand that as far as the axioms are concerned, it does not matter whether a number is written as an exponent, a radical, a logarithm, a sine, a cosine, or in any of the many ways in which a number can be written, what matters is that it is a number—everything, anywhere in every equation, which is a number, is governed by the axioms and definitions of algebra.

Chapter 5 Exercises

1. Prove that $\sqrt{3}$ is irrational. (☆12) *Hint: proceed as was done with $\sqrt{2}$, only this time show that the numerator and denominator must both be odd, then show that 3 is a common factor for both the numerator and denominator – a contradiction – by considering the remainders $(0, 1$ and $2)$ of an odd number whenever it is divided by 3 —use your intuition (this problem is not for the faint hearted).*

2. Perform the multiplication or addition using only the rules of radicals and simplify (i.e., make it easier to read).

   a. $\sqrt[7]{16j^3}\cdot\sqrt[7]{8j^6}\cdot\sqrt[7]{j}$ b. $(3-\sqrt{6x^3})(\sqrt{6x^3}+3)$ c. $2\sqrt[3]{x^2}+\sqrt[3]{x^2}$

   d. $\sqrt[3]{x-1}\cdot\sqrt[3]{1-x}\cdot\sqrt[3]{x-1}$ e. $5x\sqrt[3]{4xy}+3\sqrt[3]{2x^2}\cdot\sqrt[3]{2x^2y}$

   f. $\sqrt[5]{-486x^2y^6}$

3. Perform the multiplication using either the rules of radicals or exponents and simplify whenever possible. (*Hint: it may be easier to first transform radical notation into exponential notation before calculating and then determine the least common denominator among all exponents.*)

   a. $\sqrt{3x}\cdot\sqrt[3]{4y^2}$ b. $\sqrt{k}\cdot\sqrt[3]{k^2}\cdot\sqrt[6]{k^5}$ c. $\sqrt[5]{a^2b^3y}\cdot\sqrt[3]{ab^2x}\cdot\sqrt[15]{a^4b^{11}y^{12}}$

   d. $(jk)^{1/2}\cdot(jk^3)^{1/5}\cdot(j^2k)^{3/7}$ e. $\sqrt{x^2-1}\cdot\sqrt{x^2+x}\cdot\sqrt{x}$

   f. $\sqrt{\sqrt{x-5}}\cdot\sqrt{\sqrt{x+5}}$

4. Prove that a rational number times an irrational number is irrational. (*Hint: assume that all Real numbers must be one or the other and nothing "in between", then prove it by contradiction.*)

5. Approximate the value of the Golden Ratio $(1.618\ldots)$ to two decimal places by approximating the value of $1/b$ in the equation $1+b=1/b$. (*Hint: $b=1$ is too large and $b=0.5$ is too small, so, keep halving or guessing the best distance between the numbers until you get to the desired decimal approximation.*)

6. Prove that $\sqrt{a+b}=\sqrt{a}+\sqrt{b}$ if and only if $a=0$ or $b=0$ (or both).

7. Simplify and "eliminate" negative exponents, and if appropriate, write your results in radical notation.

   a. $(x^{1/2}y^{5/2})^{2/3}$ b. $\left(\dfrac{y^6}{9a^{-4}}\right)^{1/2}$ c. $b^{3/4}(b^{2/3}+b^{4/3})$ d. $\left(\dfrac{3y^{-6}}{[9a^{-4}+6b^{-4}]^{-1}}\right)^{-1/2}$

8. Verify that when working with powers of powers of powers, etc., one must work from the topmost power (literally speaking), down to the bottom most power. In other words, show that

$$2^{5^{3^2}}=2^{5^{(3^2)}}\neq 2^{(5^3)^2}$$

   and explain why the right-hand side of $\neq$ is incorrect for what's stated.

9. Simplify the following powers of powers. (When in doubt write it *all* out.)

   a. $2^{2^{2^2}}$ b. $(3y)^{9^{(1/2)}}$ c. $(2z)^{16^{(1/2)^3}}$

Chapter 3 Exercises

CHAPTER 6
CONTINUED STUDY

Every axiom and every definition in this book were needed to identically construct the ancient (and modern) history of elementary arithmetic. Remove anyone of them and the construction fails, but that has not stopped mathematicians from examining algebraic structures which have fewer or different axioms. Abstract Algebra, which deals with the concepts known as groups (which we have already encountered), fields and rings, among other algebraic concepts is the study of all such mathematical structures and their relation to analogous (isomorphic) forms.

In fact, there are many types of "arithmetics" that abide by fewer axioms than those presented here. The arithmetic of matrices, for example, which one learns about in Algebra 2, does not abide by the commutative property of multiplication—changing the order of the matrix-factors changes the product.

Although we have considered all of the axioms of elementary algebra (with, perhaps, the exception of the induction axiom) it is only a very small fraction of what algebra is and what it can do. There are many more definitions to consider, more methods available to analyze equations, and more questions that need to be answered if our understanding of algebra is to advance.

6.1 LOGARITHMS

For example, we may ask, "what is the value of 2^{π} ?" This seems a reasonable question to ask even though π is an irrational number because the value of π ($=3.1415926...$) approximately equals the fraction $22/7$ (≈ 3.14286). Thus,

$$2^{\pi} \approx 2^{22/7} = \sqrt[7]{2^{22}} = \sqrt[7]{4193404} \approx 8.832716$$

We could get an even better approximation of 2^{π} using the fraction $355/113 \approx 3.141593$. Thus,

$$2^{\pi} \approx 2^{355/113} = \sqrt[113]{2^{355}} \approx 8.824979\,.$$

By considering better and better fractional approximations to π, we should be able to get as close to the value of 2^{π} as we like. We could, of course, just as easily ask the question of π^{π}, which, from all that we've said, would approximately equal

$$\sqrt[113]{(355/113)^{355}} \approx 36.4622\,.$$

Before the advent of calculators these calculations were done with nothing more than pen and paper using anything from geometry—as was the case for the Ancient Greek mathematician Archimedes—to calculus (and beyond), which was first introduced by Newton and Leibnitz in the middle to late 17th century. In fact, mathematicians of earlier times laboriously compiled huge volumes of hand calculated tables for quick reference. Many of the earlier tables had the calculated values for trigonometric functions, the calculations of which otherwise (if they had to be done each time) would have required using Pythagoras' theorem repeatedly.

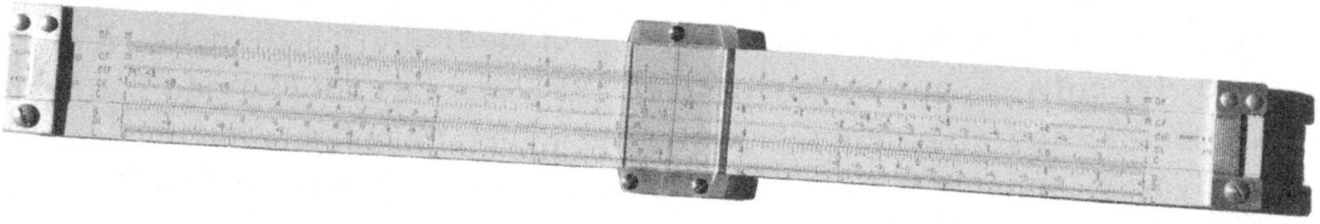

Figure 26: A slide rule. The hatch marks were spaced logarithmically – not linearly (i.e., not equally) apart. Their precision and accuracy were linked to their size. They are now mostly just collector items.

Slide rulers (Figure 26) and logarithmic tables (Table 9), which have long since been replaced with calculators, allowed scientists and mathematicians of the past to quickly calculate products and quotients by reducing such operations to addition and subtraction. They also allowed mathematicians to approximate the value of numbers like 2^{π}, π^{π} or $\sqrt[3]{7}$ by reducing exponential operations to the simpler operations of multiplication and division.

Unlike other mathematical inventions (discoveries?), which are usually preceded by a century or so of intense labor and debate among many mathematicians, the invention of logarithms seems to be the singular creation of John Napier (circa 1614), who was the first to ever write about them. In modern times logarithms are used mostly as decision subroutines in computer science; to describe such things as sound intensity, acidity, and steam pressure; and to estimate the size of earthquakes and volcanic eruptions.

From a mathematical perspective, logarithms are important because they undo what (base) numbers do to their exponents. You already know how to undue the cubing of a number by taking its cube root, i.e., $\sqrt[3]{x^3}=x$; but how does one undo 2^x to get back x? Well, that's where logarithms come in because logarithms are defined in such a way that $\log_2(2^x)=x$. Looked at another way, if $2^3=8$, then $\log_2 8=3$. In English we would say *if the base number* 2 *raised to the* 3rd *power is* 8, *then the logarithm of* 8 *to base* 2 *is the power/exponent* 3. In other words, logarithms answer questions like, "to what power must 2 be raised to get 8?" The answer is 3; specifically, $\log_2 8=3$.

○ DEF. 6.1.1: LOGARITHM
For $x, b>0$ *and* $b\neq 1$, $\log_b x=y$ *if and only if* $b^y=x$

Logarithmic tables (and slide rules), which were popular right up until the mid-1970s, can be constructed for any base number. The only problem is that there are so many numbers to choose from (infinitely many) that one wouldn't know where to start. So, bygone mathematicians had decided to create logarithmic tables for just two bases: base-10 and base $e=2.71828...$ (the, so called, natural logarithm*). Table 9 below shows an excerpt from a base-10 logarithmic table—base-10 logarithms are also known as common logarithms.

To understand how logarithms could be used to aide in multiplication and division, we will multiply the number 2.03 by 1.67 using logarithms: Looking at Table 9 (below), we would first look up the logarithms for each of these numbers, which according to the table are 0.3075 and 0.2227, respectively. Upon adding these numbers, we get 0.5302. Now, working in reverse, we must look for the number whose logarithm is this number (called the anti-logarithm), and that number is 3.39, which is not a bad approximation. The exact answer, using a calculator, is 3.3901.

The reason this works, if you recall, is because when we multiply two or more identical base-numbers each raised to an exponent, we add the exponents, specifically, $a^x \cdot a^y = a^{x+y}$. In our example we have,

$$(10^{0.3075})\cdot(10^{0.2227}) = 10^{0.3075+0.2227} = 10^{0.5302} \quad \Leftrightarrow \quad 2.03\times 1.67 = 3.3901$$

The equations on either side of the symbol $\Leftrightarrow$ say the exact same thing because $10^{0.3075} \approx 2.03$, $10^{0.2227} \approx 1.67$ and $10^{0.5302} \approx 3.39$.

Table of Common Logarithms—Log$_{10}x$

x	0	0.01	0.02	0.03	0.04	0.05	0.06	0.07	0.08	0.09
0.1	-1.0000	-0.9586	-0.9208	-0.8861	-0.8539	-0.8239	-0.7959	-0.7696	-0.7447	-0.7212
0.2	-0.6990	-0.6778	-0.6576	-0.6383	-0.6198	-0.6021	-0.5850	-0.5686	-0.5528	-0.5376
0.3	-0.5229	-0.5086	-0.4949	-0.4815	-0.4685	-0.4559	-0.4437	-0.4318	-0.4202	-0.4089
0.4	-0.3979	-0.3872	-0.3768	-0.3665	-0.3565	-0.3468	-0.3372	-0.3279	-0.31	.3098
	-0.3010	-0.2924	-0.2840	-0.2757	2676		-0	-0.2441		291
	59	0.1173	0.1206	0.1		1	0.	0.1335		
	0.1461	0.1492	0.1523	0.155	.584	0.1614	0.1644	0.1673		0.
1.5	0.1761	0.1790	0.1818	0.1847	0.1875	0.1903	0.1931	0.1959	0.1987	0.2014
1.6	0.2041	0.2068	0.2095	0.2122	0.2148	0.2175	0.2201	**0.2227**	0.2253	0.2279
1.7	0.2304	0.2330	0.2355	0.2380	0.2405	0.2430	0.2455	0.2480	0.2504	0.2529
1.8	0.2553	0.2577	0.2601	0.2625	0.2648	0.2672	0.2695	0.2718	0.2742	0.2765
1.9	0.2788	0.2810	0.2833	0.2856	0.2878	0.2900	0.2923	0.2945	0.2967	0.2989
2	0.3010	0.3032	0.3054	**0.3075**	0.3096	0.3118	0.3139	0.3160	0.3181	0.3201
2.1	0.3222	0.3243	0.3263	0.3284	0.3304	0.3324	0.3345	0.3365	0.3385	0.3404
2.2	0.3424	0.3444	0.346	0.3483	0.3	0.3522			0.3579	0.3598
		363								
								0.		
	0.4771	0.478	0.4	0.4814		0.4843	0.4857	0.4871	.4886	0.4900
3.1	0.4914	0.4928	0.4942	0.4955	0.4969	0.4983	0.4997	0.5011	0.5024	0.5038
3.2	0.5051	0.5065	0.5079	0.5092	0.5105	0.5119	0.5132	0.5145	0.5159	0.5172
3.3	0.5185	0.5198	0.5211	0.5224	0.5237	0.5250	0.5263	0.5276	0.5289	**0.5302**
	15	0.5328	0.534	0.5353	0.5366	0.5378	0.5391	0.5403	0.5416	0.5428

Table 9: Excerpt from base 10 (common) logarithmic table

Let's take a closer look at the equation $10^{0.5302} \approx 3.39$ from the preceding paragraph. The number 0.5302 in the exponent position is the logarithm; the number 10 is the base; and the number 3.39 is the number whose logarithm is 0.5302. When the equation is written as $10^{0.5302} \approx 3.39$ it is easy to glean the

* The number e shows up in many natural processes, much like the Golden Ratio.

logarithm of 3.39 from the exponent of 10 (i.e., 0.5302). However, the logarithm of a number is usually not known in advance, hence we use the short-hand notation, (in this example) $\log_{10}3.39$, to describe it; thus, the number $\log_{10}3.39$*, which equals 0.5302, is not only a logarithm it is also an exponent.

Because logarithms are exponents, we can use the rules of exponents from the previous chapter when manipulating them. One such rule is the following: *the logarithm of a product is equal to the sum of the logarithms of the factors that make up the product.* In short-hand we would write,

<u>Rule No. 1</u>: $\log_b(ac) = \log_b a + \log_b c$.

Thus, for example, $\log_{10}35 = \log_{10}(5 \cdot 7) = \log_{10}5 + \log_{10}7$. Using a calculator, one can easily test the validity of this equation. Furthermore, it is important to note that $\log_{10}a^2 = \log_{10}(a \cdot a) = \log_{10}a + \log_{10}a = 2\log_{10}a$. In other words, *the logarithm of the square of a number is equal to two times the logarithm of the number.*

This brings us to the next two general rules regarding logarithms (although the one implies the other): *the logarithm of an exponentiated number is equal to the exponent of the number times the logarithm of the number*; in symbols we have,

<u>Rule No. 2</u>: $\log_b a^x = x\log_b a$; and <u>Rule No. 3</u>: $\log_b a^{-x} = -x\log_b a$.

Thus, $\log 8^7 = 7\log 8$, $\ln 2^\pi = \pi \cdot \ln 2$ ($\ln x \equiv \log_e x$ †) and $\log \pi^{-\pi} = -\pi \cdot \log \pi$. If we combine Rule No. 1 with Rule No. 3, we have $\log_b(a/c) = \log_b(a \cdot (1/c)) = \log_b(a \cdot c^{-1}) = \log_b a - \log_b c$. For example, $\log_{12}(4/5) = \log_{12}4 - \log_{12}5$.

Their proofs depend on the fact that their exits a logarithm for every positive number for *any* positive ($\neq 1$) base, which we will just assume to be true.

◆ THEOREM 6.1.1

Rule No. 1: $\log_b(ac) = \log_b a + \log_b c$

Proof: let $\log_b a = j$ and $\log_b c = k$, then, by the definition of logarithms (Def. 6.1.1), we have $b^j = a$ and $b^k = c$, respectively. Therefore, $ac = b^j b^k$. However, $ac = b^j b^k = b^{j+k}$, but from the definition of logarithms (Def.6.1.1) this equation yields back $\log_b(ac) = j + k$. Substituting for what j and k equals, we have the desired result: $\log_b(ac) = \log_b a + \log_b c$.

◆ THEOREM 6.1.2

Rule No. 2: $\log_b a^x = x\log_b a$

Proof: let $\log_b a = j$, then $b^j = a$. Raising both sides of this equation by x yields $(b^j)^x = a^x$ or equivalently $b^{xj} = a^x$, which by definition means $\log_b a^x = xj$. Substituting for what j equals, we have the desired result: $\log_b a^x = x\log_b a$.

---

* When working with common logarithms, the base number, 10, is usually omitted; thus, $\log_{10}a = \log a$. (When working with natural logarithms, we define $\ln a = \log_e a$ ($e = 2.718\ldots$).

† The symbol $\equiv$ is an abbreviation which means "is defined as" or "is defined to be".

◆ **THEOREM 6.1.3**

Rule No. 3: $\log_b a^{-x} = -x\log_b a$ (This follows directly from Theorem 6.1.2.)

Finally, logarithms of numbers which are less than or equal to zero are not defined (although we can get as close to zero—from the positive side—as we like); in other words, $\log_b a$ is defined provided both a and b are greater than zero (and $b \neq 1$; why?).

To see why this is necessary suppose we have the equation $\log_3(-27) = x$, then, recalling how this equation is defined, we have $3^x = -27$. As you can see, we can find no real-number value of x which will make this equation true: $x = 3$ results in a *positive* 27, while $x = -3$ results in the fraction $1/27$.

On the other hand, a logarithmic equation with a negative base behaves like a jittery light bulb ready to burn out: at some positive numbers it will pop into existence while at others it will pop out (i.e., be undefined or imaginary). Recall again that $y = \log_{(-3)} x$ if and only if $x = (-3)^y$; thus, for example, the equation is defined when $y = 1/11$ because there exists a (an exact) number $x \approx -1.105$ that when raised to the 11th power equals -3. However, when $y = 1/12$ no such real-number x exists; it exits again when $y = 1/13$ and fails again when $y = 1/14$, etc. Hence, it is agreed that base numbers for logarithms are always greater than zero. (The problems with $\log_b 0$, $\log_1 a$ and $\log_0 a$ are for you to explore.)

~

6.2 POLYNOMIAL EQUATIONS AND FACTORING

We may also wonder what values of x $(x \in R)$ will make equations of the sort $3x^2 - 9x - 12 = 0$ or $3x^3 + 2x^2 - 3x - 2 \leq 0$ true, or what values of x will make them both true at the same time, or whether any solutions exist at all. Problems like these involve finding the *roots* (i.e., solutions) of *polynomial* equations*, which often require some very clever methods for solving. Some of these methods had been toiled over for years before mathematicians finally arrived at a universally accepted method. Fortunately for us, we can build new knowledge on top of old without all the sweat and tears, but the penalty is that it requires considerable memorization and maintaining some level of activity in mathematics throughout your entire life.

To get an idea of how something like $3x^2 - 9x - 12 = 0$ can be solved, we might consider converting this addition problem into one of multiplication. If we can do that, then we can use the theorem which states that if the product of two numbers is zero, then one or the other of the numbers must be zero (Theorem 4.5.2). Let's see how that would work.

---

* A polynomial is an expression involving any number of terms, each of which involve the products of numeral numbers and variables raised to non-negative exponents. For example, $3yx^7 - 2w^2$ is a binomial, $3x^2 - 9x - 12$ is a trinomial and $4x^5 + 17x^4 - x^3 - 8x^2 + x + 3x^0$ is a polynomial. And, of course, a monomial is never a polynomial.

First, we can factor out a 3 from this equation (distributive property) to obtain $(\underline{3}\cdot x^2-\underline{3}\cdot 3x-\underline{3}\cdot 4)=3(x^2-3x-4)=0$. Because this number is the product of two numbers, which equals zero, Theorem 4.5.2 tells us that either $3=0$ or $x^2-3x-4=0$. Obviously, $3\neq 0$; hence, it must be the case that $x^2-3x-4=0$. Now, we could certainly try to guess the number x that would make this equation true or we could try to determine what are the factors of x^2-3x-4. Doing the latter would allow us to use Theorem 4.5.2 again. However, finding the factors of a polynomial equation is seldom easy. It requires practice and it always involves the distributive property. Initially, however, what is important is learning a general method for finding factors—first things first.

Begin by setting up two pairs of empty parentheses to signify the product of two (yet unknown) numbers. We suspect that it is not the product of three or more numbers because the term x^2 is the product of two numbers: $x\cdot x$.

$$\textit{Step 1: } (\qquad)(\qquad)=0.$$

Because one of terms (the first) in the expression x^2-3x-4 is x^2 ($=x\cdot x$), we can put an x inside each of the empty parentheses, thus ensuring that when we carry out the multiplication we will retrieve back the term x^2:

$$\textit{Step 2: } (x\qquad)(x\qquad)=0.$$

Because there is the integer -4 in the expression, there must be two numeral-numbers (although practically speaking it is usually not so simple), one for each of the parentheses, whose product is -4, and it must be the case that one of the numbers is positive and the other negative. The pairwise factors of -4 are $(-1)\cdot 4$, $1\cdot(-4)$ and $2\cdot(-2)$. Observing that the only way we can make a sum from these factors that equals -3 (the middle term of the expression x^2-3x-4 [the "x" is not important in our consideration]) is with either the factors (-1), 4 or with the factors 1, (-4), we select the latter pair because we need more negative numbers than positive numbers in the sum. Thus, we place a factor in each of the parentheses as follows:

$$\textit{Step 3: } (x\qquad +1)(x\qquad -4)=0;$$

finally, we write $(x+1)(x-4)=0$. We can check this result by performing the multiplication (using the distributive property) and see if we arrive at our original expression: x^2-3x-4.

Having found the product of two numbers which equals 0, we may now use Theorem 4.5.2 to deduce that either $(x+1)=0$ or $(x-4)=0$, and that consequently, either $x=-1$ or $x=4$. In other words, both of these values for x (each at a time) will make the equation $3x^2-9x-12=0$ true (but don't take my word for it—check it).

To see how we could solve polynomial inequalities, say $3x^3+2x^2-3x-2\leq 0$, we would once again try to transform this addition problem into a multiplication problem (all the useful theorems seem to involve multiplication), the reason for which we will see momentarily. Again, to know when and how to do this requires some familiarity with such problems, but we must begin somewhere. In the following four transformations of the original inequality it is important to be aware of how the distributive axiom is being used. Convince yourself that they are indeed equivalent to the original inequality (i.e., are equivalent to $3x^3+2x^2-3x-2\le 0$).

1. $(3x^3+2x^2)-(3x+2)\le 0$
2. $x^2(3x+2)-1\cdot(3x+2)\le 0$
3. $(x^2-1)(3x+2)\le 0$
4. $(x+1)(x-1)(3x+2)\le 0$

The last result was arrived at by using the method discussed in the previous example involving equality. Furthermore, whatever can be said about this last inequality can be said about the original inequality.

But what does it say? It says that the product of three numbers is less than or equal to zero. Therefore, it cannot be the case that all three factors are greater than zero because the product of three numbers greater than zero is a number greater than zero. It can also not be the case that two of the factors are less than zero and one of them greater than zero because the product of two numbers which are less than zero is a number that is greater than zero, and, again, the product of two numbers each greater than zero is a number greater than zero. That leaves us with only the following possibilities.

Case 1. One of the factors is less than zero and the other two are greater than zero. Thus,

a. $x+1<0$ and the other two factors are greater than zero.

b. $x-1<0$ and the other two factors are greater than zero.

c. $3x+2<0$ and the other two factors are greater than zero.

Case 2. All three factors are less than zero.

Case 3. One or the other of the three factors equals zero (to satisfy the condition when the product of all three numbers equals zero).

Thus, to evaluate which of these cases will make $3x^3+2x^2-3x-2\le 0$ true, we need to determine which of these cases do not lead us into a contradiction.

Case 1a: $x+1<0$ *and* $x-1>0$ *and* $3x+2>0$. This implies that $x<-1$ and $x>1$ and $x>-2/3$; but this is a contradiction because a number, x, cannot be less than -1 and simultaneously greater than 1 (the last inequality doesn't need to come into play). In set interval notation we would write $(-\infty,\ -1)\cap(1,\ \infty)=\varnothing$. <u>Case 1a: contradiction</u>

Case 1b: $x-1<0$ *and* $x+1>0$ *and* $3x+2>0$. This implies that $x<1$ and $x>-1$ and $x>-2/3$. All three of these inequalities *can* be true simultaneously, $x=1/2$ is an example: it is less than 1, greater than -1, and greater than $-2/3$. However, we need to find *all* the numbers for which this equation is true. Using set interval notation, we would write $(-\infty, 1)\cap(-1, \infty)\cap(-2/3, \infty)=(-2/3, 1)$. Thus, every number, x, in the open interval $(-2/3, 1)$ is a solution. Case 1b: solutions exist

Case 1c: $3x+2<0$ and $x-1>0$, and we can stop right there because these two inequalities imply that $x<-3/2$ and $x>1$: no such number exists. Case 1c: contradiction

Case 2: All three factors are less than zero. In that case $x<1$, $x<-1$, and $x<-2/3$. All three of these inequalities will be satisfied whenever $x<-1$,* i.e., whenever $x\in(-\infty, -1)$. Case 2: solutions exist

Case 3: $(x+1)(x-1)(3x+2)=0$. This will be true whenever $x=-1, 1$ or $-2/3$. (*x can only be one number at a time and none of these numbers contradict the equation, try substituting each for x.*) Case 3: solutions exist

Thus, the solutions to $3x^3+2x^2-3x-2\leq 0$ are represented by the union of all the solution sets; therefore, $(-2/3, 1)\cup(-\infty, -1)\cup\{1\}\cup\{-1\}\cup\{-2/3\}=(-\infty, -1]\cup[-2/3, 1]$ is the solution set. (Thus, for example, $-3/4$, which falls between these two intervals, *is not* in the solution set.)

Of course, there is always another difficult problem waiting to be undertaken and new, creative mathematics waiting to be written, but without some persistence and some logical, theoretical, framework from which to approach these problems they will forever remain nothing more than idle thoughts.

~

6.3 GRAPHING

In the seventeenth century René Descartes came upon the clever idea of using a pair of perpendicular lines—*axes*—to visualize the values of a two-variable equation (Figure 27), thereby creating another method for finding the solutions to equations like those just discussed.

For example, suppose we wanted to find the solution set to $x^2-1.5=0$. Although this is a rather simple example, the principle behind the method remains the same. From the equation $x^2-1.5=0$, we can write the more general equation $y=x^2-1.5$ (it is customary to put the independent variable to the left of equals and the dependent variable to the right). Our original equation, $x^2-1.5=0$, could then be represented by $y=0$. Writing the equation in this fashion, we are then (1) able to choose various values for x and determine the corresponding y value; (2) create what is called a table of values (Table 10); and (3) plot the values on a two dimensional graph; thus, enabling us to visualize, approximately, where the y values drop closer and closer to zero and locate about where that happens in relation to the x-axis, voilà, solutions found—even if only approximate. (What is the exact solution for $y=x^2-1.5$?)

* All three inequalities will be satisfied whenever $x<-772$ too; however, we would have missed all of the numbers from -772 up to and including -1; the goal is to find *all* solutions.

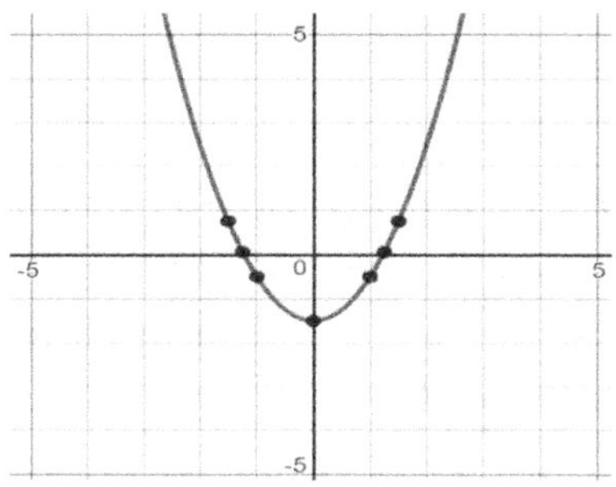

Figure 27: The visualization (graph) of $y = x^2 - 1.5$ using perpendicular axes (lines), x and y.

x	0	1	−1	1.5	−1.5	1.25	−1.25
y	−1.5	−0.5	−0.5	0.75	0.75	0.0625	0.0625

Table 10: A short table of values for the equation $y = x^2 - 1.5$

Graphing equations also makes it easier for us to find and understand the solution set for groups of equations, which, except for *set builder notation,* which we will discuss momentarily, would be difficult to represent otherwise.

Consider the two inequalities $y<(x+1)^2$ and $y \le -(x/2)+3$, and suppose we wish to find every pair of values for x and y, which we will label as the *ordered pair* or *point* (x,y) (the x value always comes first), that will make both equations true simultaneously. For example, the ordered pair $(1,2)$, shown in Figure 28, is a solution to both equations. We can test it: substitute 1 for x and 2 for y in each equation and show that each inequality is true.

On the other hand, the ordered pairs $(-4,6)$ and $(-2,3)$ are solutions for one or the other of the two inequalities, but not both, and the ordered pair $(0,5)$ is not a solution to either inequality—it makes both inequalities false. Thus, we must choose y values that are *below* both graphs (because *down* means less relative to the y-axis [while *left* means less relative to the x-axis]) to ensure that both inequalities yield true statements. The overlapped, *darkest* shaded area represents all those ordered pairs that are solutions to both inequalities. (Note also that the graph for $y<(x+1)^2$ is represented with a dashed line—because points on that line (e.g., the point $(1,4)$) would yield false statements.)

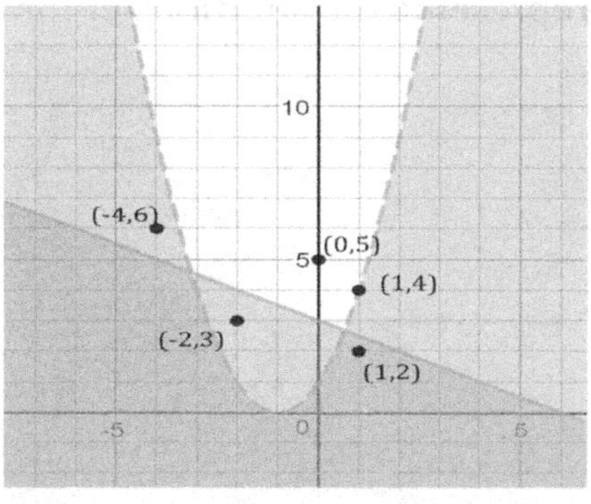

Figure 28: Graph of the solution set for two simultaneous inequalities.

One can also describe the solution set using set builder notation, which is nothing more than the swapping out of English for symbols. Thus, to describe the solution set for the two inequalities we've just examined, we would write,

$$\{(x,y) | y<(x+1)^2 \wedge y \leq -x/2+3\}.$$

In English it reads, *the set of all ordered pairs x-y such that y is less than the square of x plus one and y is less than or equal to one-half the additive inverse (or the negative) of x plus three.* The symbol " | " takes the place of the words *such that* and the symbol $\wedge$ takes the place of the word *and.*

Now, you must admit that this notation describes precisely the solution set to these two inequalities, the only problem is that we are no better off naming what those ordered values for x and y are than when we started, and there is no better way to explicitly exhibit the solution set then with the picture already presented (Figure 28). However, this notation is somewhat better when describing other solution sets. For example, to describe the solution set for the equation $3x^3+2x^2-3x-2 \leq 0$ we would write $\{x | x \in (-\infty, -1] \cup [-2/3, 1]\}$, which reads, *the set of all x such that x is an element of the right-closed interval from negative infinity to negative one union the closed interval from negative two-thirds to one.* (It is clearly apparent that math written entirely in English would use many times the paper space—but it can be done.)

While we are on the subject of inequalities, it is important to recall Axiom O4: *if $a<b$ and $0<c$, then $ac<bc$.* It states that in order to multiply both sides of the inequality $a<b$ by c, then c must be greater than 0. Therefore, if we wish to isolate the independent variable x in the rather simple inequality

$$\frac{y}{x}<2$$

by multiplying both sides of the inequality by x —the multiplicative inverse of $1/x$ —then we must know whether x is greater or less than 0 (in this equation x cannot equal 0, but we can choose values of x as close to 0 as we like). Thus, we have two cases to evaluate: $0<x$ (i.e., $x>0$) and $x<0$.

Case 1: $0<x$. In this case Axiom O4 applies and we may freely write $y<2x$. The graph of this pair of inequalities immediately follows (the *darkest* shaded area represents the solution set for this case):

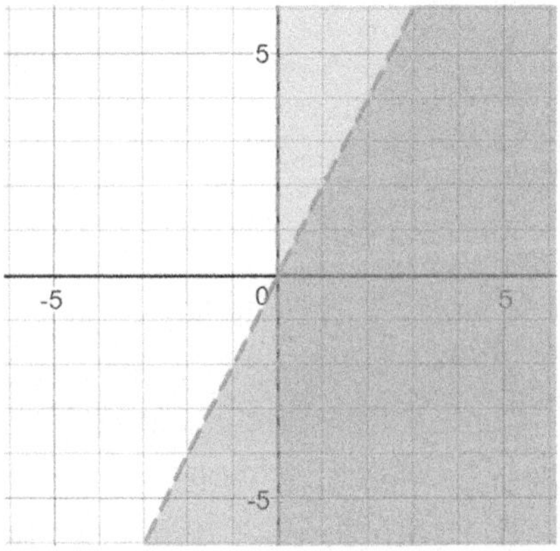

Figure 29: the graph of y<2x and 0<x (the darkest region)

Case 2: $x<0$. In this case Axiom O4 does not apply. However, $x<0$ implies (adding $-x$ to both sides: Axiom O3) $0<-x$, which does apply. Thus, we may write (Axiom O4)

$$(-x)\frac{y}{x}<(-x)\cdot 2\text{, which implies } -y<-2x,$$

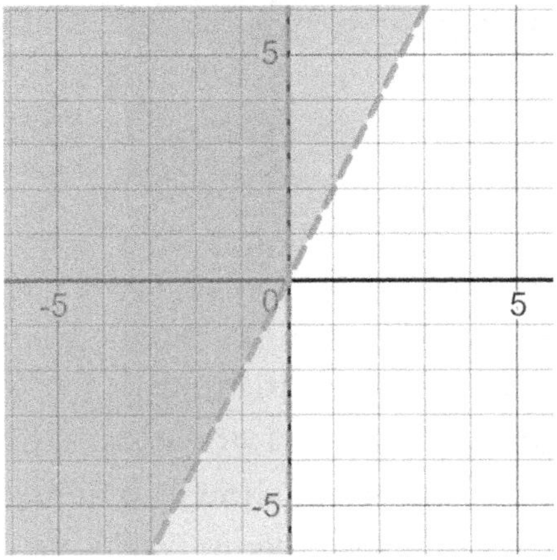

Figure 30: the graph of y>2x and x<0 (darkest region)

from which we may now apply Axiom O3 to both sides by adding y and $2x$ to both sides (additive inverses), upon which we obtain $y>2x$. The graph of this pair of inequalities (i.e., $x<0$ and $y>2x$) is shown in Figure 30 (again, the darkest shaded region):

Therefore, the graph of the solution set for the equation

$$\frac{y}{x}<2$$

is the union of the darkest regions from Figure 29 and Figure 30, which is represented by *all* the shaded regions in Figure 31.

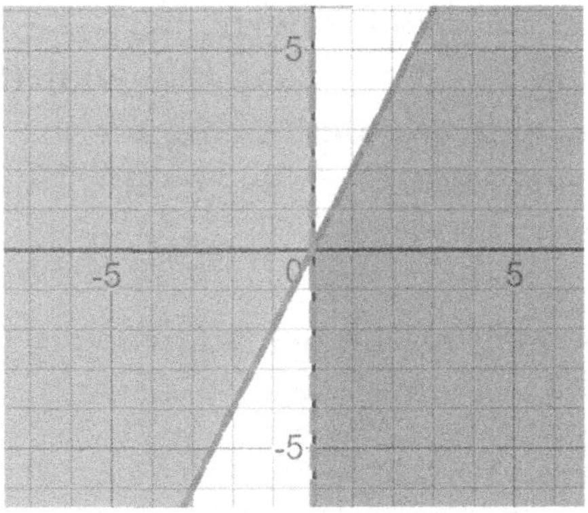

Figure 31: the solution set for the inequality y/x<2 (the two shaded regions)

On the other hand, if we had started the process of finding the solution set of the inequality $y<2x$ instead of $(y/x)<2$, the solution set would have been completely different. The solution set to $y<2x$ is shown in figure 32:

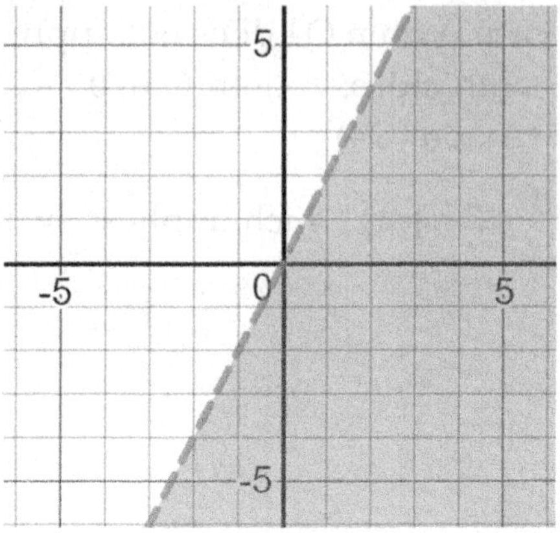

Figure 32: the solution set for the inequality y<2x (shaded region)

The reason for the difference rests with the inequality we start with, which must be abided. If we start with the inequality $(y/x)<2$, then we wind up with the solution set shown in Figure 31; if we start with the inequality $y<2x$, then we wind up with the solution set shown in Figure 32.

If we intuitively analyze the two inequalities, we can see how this makes sense. For the inequality $(y/x)<2$ we can choose negative or positive values for x, followed by positive or negative values, respectively, for y or the other way around provided one number is positive and the other is negative (although there are other solutions in which both numbers carry the same sign). In other words, the order of our choices is not important, the result will always be a number less than some positive number, like the number 2. On the other hand, for the inequality $y<2x$, whenever we choose a negative value for x we must choose a negative value for y, and whenever we choose a positive value for y we must choose a positive value for x; otherwise we run into a contradiction. For example, the points $(-5,5)$ and $(-5,-5)$ are solutions to $(y/x)<2$, but not solutions to $y<2x$, while the point $(-1,-5)$ is not a solution to $(y/x)<2$, but is a solution to $y<2x$, and the point $(1,5)$ is not a solution to either inequality. Remember, we are looking for the solution set of ordered pairs which relates to the *initially given* inequality. Thus,

$$\{(x,y)|\frac{y}{x}<2\} \neq \{(x,y)|y<2x\},$$

even though the two inequalities are related in a very fundamental way.

~

6.4 FUNCTIONS

In the late seventeenth and early eighteenth century, the idea of functions was also beginning to take root, owing especially to the works of mathematicians like Newton and Leibnitz in calculus. The idea of functions found its way into almost every branch of mathematics, and algebra is no exception. The definition of a function is rather simple to state and understand:

○ DEF. 6.4.1: FUNCTION

A function from a set A to a set B is a correspondence that relates each element of set A to exactly one element of set B (see Figure 33 for an example).

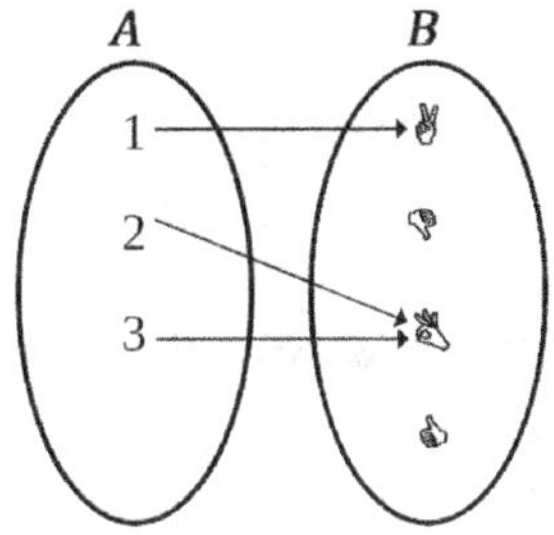

Figure 33: A function from set A to set B (Does it abide by the definition?).

However, when this definition is viewed in all its logical profundity it can produce some very strange, complex mathematical objects. An example of the logical complexity that arises from this seemingly simple definition is the following function, known as the Popcorn Function:

$$y=\begin{cases} 0 & \text{if } x \text{ is irrational} \\ \frac{1}{n} & \text{if } x=\frac{q}{n} \text{ where } n\in Z_P\text{ , } q\in Z \text{ and } \frac{q}{n} \text{ is in lowest terms} \end{cases}$$

(The left brace "{" in the function definition should not be mistaken for the notion of a set. A set requires a left and a right curly brace, "{, }". The use of the left brace merely denotes that y can take on either value, 0 or $\frac{1}{n}$, based upon the conditions which apply.)

Before we get into a few of the details of the complex graph of this function, a few comments about the function itself are in order. First, the function indicates that x takes on every real number from negative infinity to positive infinity. For example, if $x=-\sqrt[3]{102{,}405}$ or $\pi^{7{,}690}$, both of which are irrational numbers in the "interval" $(-\infty,\infty)$, then $y=0$. Likewise if $x=990/980$, $2/196$, $-513{,}783/98$ or $1/98$, then when these fractions are reduced to their lowest-terms fraction (if not already) we have $y=1/98$ for each of these x values. Second, the y values are such that $0\leq y\leq 1$.

Figure 34 (on the reverse) shows the graph of the, aptly named, Popcorn Function. It repeats itself on every unit interval from negative infinity to positive infinity. The graph is also discontinuous (i.e., points are not connected to one another in a continuous way) at every rational number, and continuous at every irrational number. For example, the rational points at (1,1), $(\frac{1}{2},\frac{1}{2})$, $(\frac{3}{2},\frac{1}{2})$, $(\frac{1}{3},\frac{1}{3})$, $(\frac{2}{3},\frac{1}{3})$, etc., do not form a continuous line with their neighbors—they are just isolated points "floating" in a Cartesian coordinate plane. However, showing that the rational numbers are discontinuous, and the irrationals continuous is beyond the scope of this book. That aside, this graph is a direct, logical consequence of the accepted definition of a function and the principles elaborated in modern set theory.

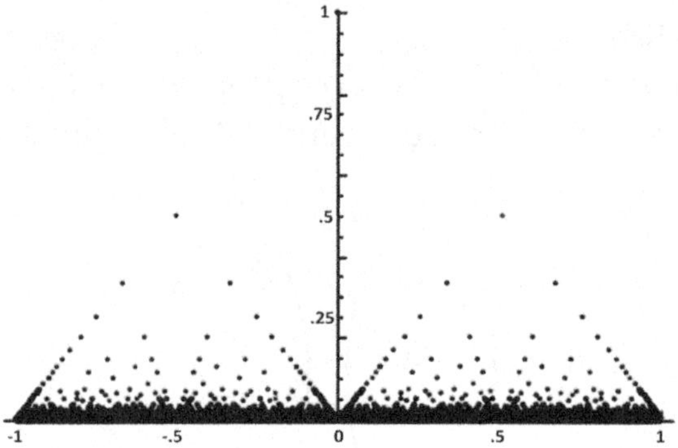

Figure 34: Graph of the Popcorn Function.

It almost seems as if Carl Thomae (circa 1900), who thought up this function, went out of his way to provoke an argument about the accepted definition of a function, which is likely what he intended to do. So, what are mathematicians to do? Should they revise the definition of a function and rethink the theory of sets to eliminate such weird concoctions? However, if they did, they would have to deal with whatever strange monsters could be lurking in the revisions. It is sometimes best to leave theory as it is and let logic lead where it will.

It is also possible—desirable, in fact—to have functions within functions, which in mathematical jargon are called *composite* functions. So, for example, if $z=\log_7 y$, $y=5x^2-a$ and $x=3t+1$, then, upon making the substitutions for y and x, we have $z=\log_7(5[3t+1]^2-a)$. In this case, the value of z depends on the value of y, but the value of y depends on the value of x, and the value of x depends on the value of t. In other words, within each function there can be—we can have—another function. In this regard, the composition of functions can be likened to Matryoshka nesting dolls (Figure 35).

Figure 35: Matryoshka nesting dolls and the composition of functions.

6.5 IMAGINARY/COMPLEX NUMBERS

Complex numbers are not particularly difficult to understand, we simply ask the question, "What number, if such a number even exists, would make the equation $a^2 = -1$ true?" We have already shown with the irrational numbers that the closure axiom of multiplication does not allow us to conclude that a is a number even if $a \cdot a$ is a number. However, if we assume that such equations mean something, then we must devise and introduce a hitherto unknown number to our already crowded set of real numbers, not to mention all the logical consequences that such an introduction would entail. Of course, the numbers of which I speak are the imaginary numbers—more commonly known as complex numbers—which involve the symbol $i = \sqrt{-1}$. The set of complex numbers, which is closed under both operations of addition and multiplication, marks the end of all the numbers needed to complete the study of algebra.

It took many centuries to convince the mathematical community that complex numbers should be included along with every other mathematical object then known to be a number. Although a good deal of the hard work involving imaginary numbers (even though he disliked them and considered them useless) can be contributed to Girolamo Cardano, who in 1545 wrote a book on the subject, the important contributions of many other mathematicians cannot be underestimated. However, it was not until Carl Gauss took up the banner for imaginary numbers in the early 19th century that imaginary numbers began to gain wide acceptance in the mathematical community.

There are three aspects about imaginary numbers that need to be respected and that is (1) the symbol i is purely algebraic (i.e., it is not a numeral, per se), thus, we "manipulate" it like we would any other algebraic symbol; (2) imaginary numbers do not abide by the multiplication rules of radicals with even indices or the multiplication of numbers whose exponents have an even number in the denominator.* In other words, $\sqrt{-16} \cdot \sqrt{-25} \neq \sqrt{(-16)(-25)}$; (3) if a is a positive, real number, then $\sqrt{-a} = \sqrt{a} \cdot i$; thus, $\sqrt{-16} \cdot \sqrt{-25} = (i\sqrt{16})(i\sqrt{25}) = (i \cdot i)\sqrt{16} \cdot \sqrt{25} = 20 \cdot (i \cdot i) = 20i^2 = -20$, to be perfectly clear, however, $\sqrt{-(-3)} \neq i\sqrt{-3} = i^2\sqrt{3} = -\sqrt{3}$, rather $\sqrt{-(-3)} = \sqrt{3}$.

∘ DEF. 6.5.1: IMAGINARY NUMBER RULES/DEFINITIONS

If $0 < a, b$, and n is an even, positive integer, then $\sqrt{-a} = i\sqrt{a}$† and $\sqrt[n]{(-a)(-b)} \neq \sqrt[n]{-a} \cdot \sqrt[n]{-b}$, where $i = \sqrt{-1}$; $i^0 = 1$; $i^1 = i$; $i^2 = -1$; $i^3 = -i$; and $i^4 = 1$ $(= i^0)$. Otherwise, imaginary numbers abide by every algebraic axiom and definition. (The roots of complex numbers (e.g., $\sqrt[n]{a+bi}$) are beyond the scope of this book.)

Imaginary/complex numbers are typically rendered in two parts, a real part

---

* Recall Theorem 5.1.2.

† It is customary to put the symbol *i* before a radical and a variable, but after a numeral number.

and an imaginary part, $a+bi$, where a is called the real part and b (not bi) is called the imaginary part, and where $a, b \in R$. Thus, either a or b, or both could be zero. When $b=0$, then the complex number reduces to the real number a; if $a=0$, then the number is considered *pure* imaginary.

Except for the above rules, complex numbers behave according to every other algebraic axiom/rule. Thus, for example, the sum and product of two complex numbers are calculated as follows:

$$(a+bi)+(c+di)=(a+c)+(bi+di)=(a+c)+(b+d)i$$

$$(a+bi)(c+di)=ac+(ad+bc)i+bdi^2=ac+(ad+bc)i-bd=(ac-bd)+(ad+bc)i.$$

Unlike real numbers, which can be ordered, the imaginary numbers cannot be ordered, at least not in any "natural" way. They can only be represented in a plane using a pair of axes, much like those used to represent the x and y values for a given ordered pair (x, y). In the case of imaginary numbers, however, the complex, albeit two-dimensional, plane is used merely as a device for locating inherently-unordered complex numbers, which is somewhat different than graphing a *point* in a real Cartesian plane. A point in a Cartesian plane exists in two dimensions, complex numbers do not; there is just no other way to structure them. Figure 36 plots a complex number in the complex plane.

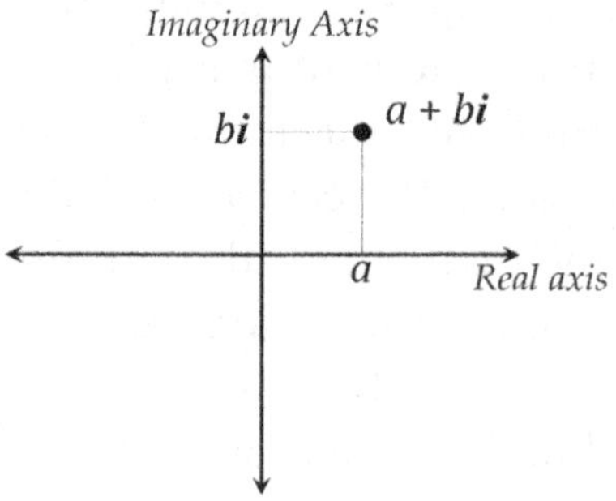

Figure 36: complex numbers represented in the complex plane

The last thing to be said, by our account anyway, about complex numbers is that there is no easy way to graph functions of complex numbers, say, for example the complex function $\{(x,y) \mid y=x^2 \wedge x,y \in \{complex\ numbers\}\}$. Functions of complex numbers "work" on two parts, the real and the imaginary parts, the result of which yields back a complex number, which is again made up of two parts. Thus, to truly graph a complex function and visualize how the dependent complex-variable, y, changes with the independent complex-variable, x, we need the assistance of four dimensions—two dimensions to visualize the dependent variable and two dimensions to visualize the independent variable. However, living in the three-dimensional world that we do presents us with a problem: how do we visualize four dimensions using only three? Although there are various techniques to understand the relation between these two complex variables, many of which involve color, and which are fascinatingly beautiful images, they all, never-the-less, have their short comings.

6.6 HIGHER MATHEMATICS

Algebra pertains to anything and everything that is considered a "number". Thus, each of the following—as single entities unto themselves—are *bound* to the rules of algebra:

$$15, x, 3xy^3+7x^2, \sqrt[8]{15z^3}-4t, \frac{\ln(3x^{2y})}{\pi^{\log z}}.$$

They are, for all intents and purposes, numbers. However, numbers come in all forms, shapes and sizes and anything that is a number, or is by name a number (but abides), may be manipulated according to the axioms of algebra.

The following two expressions are numbers in an algebraic sense:

$$\int_1^4\int_{10}^z\left[\cos^2(45\pi r^2h)\int_0^{3y}(2x^5+7)dx\right]^{e-2t}dydz, \begin{vmatrix} 3x & 2z^2 & ay \\ 15 & 9w & 2 \\ 6-b & 17 & 8 \end{vmatrix}. \quad \text{▪6.6.1}$$

Although many of you may not have had any training in trigonometry, calculus or linear algebra, that is of little concern (they all mostly involve only new number concepts or definitions anyway). What matters here is that the objects in line 6.6.1 are numbers and as such they too are bound to the rules of algebra. Thus, we may square them, cube them, take their logarithms, use them as exponents, create their multiplicative inverses (provided, of course, the expressions don't equal zero), apply the associative and commutative rules to them, and so on. In short, we can do anything with them that we could do with any other run-of-the-mill number. The only problem is that they are both rather "bulky" and time consuming to write repeatedly; so, give them a name, call them u and v and manipulate their names algebraically instead. And therein lies the tremendous power and utility of algebra.

From a historical perspective, creating (or discovering, if you prefer) the study known as algebra has been a fantastic and often tortuous struggle involving countless people over many centuries. It is truly one of the most remarkable and beautiful ideas that humans have ever conceived, and it is still not completely settled—it is still, and always will be, only a theory based on our most primitive beliefs about how numbers behave. And unlike the ease with which one can admire the beauty of a painting by Van Gogh, Rembrandt and Picasso, or appreciate the beautiful melodies of Mozart, Bach and the Beetles, the wondrous beauty of algebra is difficult to grasp, but you have made a fine go of it.

The axiomatic method is a powerful mathematical and scientific tool. Your newly acquired experience with it will serve you well in all your future math and science courses. As far as algebra goes you only need to continue practicing the axioms, theorems and definitions to fully master it and be sure to keep Appendix C at your side as you do.

Chapter 6 Exercises

1. For each of the following problems write the corresponding logarithmic equation or the corresponding exponential equation.

   a. $7^y = 49$ b. $(1/2)^3 = x$ c. $10^{4x} = 100$ d. $\log_9 81 = 2y$

   e. $\log_\pi x = y$ f. $\log_{10}(\log_\pi x) = y$ g. $e^{\ln(2x+3)} = y$

2. For each of the following problems use a scientific-calculator to calculate and compare result i with result ii with result iii (where applicable). Afterwards, convert each logarithmic problem into an exponential expression. (Recall that $\log a \equiv \log_{10} a$ and $\ln a \equiv \log_e a$.)

   a. i) $\log(5^4)^2$ ii) $2\log 5^4$ iii) $8\log 5$

   b. i) $\ln e^2$ ii) $2\ln e$ iii) 2

   c. i) $\log 10^0$ ii) $0 \cdot \log 10$

   d. i) $\log 1$ ii) $\ln 1$ iii) $\log_7 1$

   e. i) $\ln(30)$ ii) $\ln 5 + \ln(6)$ iii) $\ln 5 + \ln 2 + \ln 3$

   f. i) $\ln[(5\cdot 2)/3]$ ii) $\ln(5\cdot 2) - \ln 3$

   g. i) $\ln\sqrt[3]{(1/7)}$ ii) $(1/3)\ln(7^{-1})$ iii) $-(1/3)\ln 7$

Solving logarithmic and exponential equations is a matter of finding a common base; for example, to solve the equation $2^{3x} = 16$ we have to write the number 16 as the base 2 raised to some power. If we are familiar with the powers of 2, we would recognize that $16 = 2^4$; therefore that $2^{3x} = 2^4$. And, all things being equal, we could deduce that $3x = 4$ and hence $x = 4/3$. On the other hand, if we have an equation which no one would, normally, be familiar with, say $7^{2x+1} = \pi$, then we would need to find the exponent of 7 such that 7 raised to that exponent (i.e., $7^{\log_7 \pi}$) equals π; in other words, we need to find the value of $\log_7 \pi$. Many scientific calculators are designed to find the logarithm of any base number, for this log the result is (approx.) $\log_7 \pi = 0.588275$. Thus, $7^{2x+1} = 7^{0.588275}$, and consequently $2x + 1 = 0.588275$, and $x = -0.206$.

However, it is not necessary to use a base-7 logarithm to find this solution, because if $7^{2x+1} = \pi$ is true (which we assume it is), then $10^{7^{2x+1}} = 10^\pi$ is also true. Thus, $\log(7^{2x+1}) = \log \pi$, which implies $(2x+1)\log 7 = \log \pi$, which implies that $2x + 1 = \log \pi / \log 7$. Using our calculators to find the *base-10* logarithms of π and 7, we have $2x + 1 = 0.49715/0.8451 = 0.588275$, which yields the same result as before. We could have found the same result using any base, say base-*e*, then, $\ln(7^{2x+1}) = \ln \pi$ implies that $2x + 1 = \ln \pi / \ln 7 = 1.14473/1.94591$, which is also equal to 0.588275.

3. With that said, for each of the following exponential and logarithmic equations, solve for the unknown variable.

   a. $e^{3x-7} = e^{49}$ b. $2^{4x-6} = 64$ c. $27 = 3^{5x} \cdot 9^{x^2}$ d. $3^x = 2^{x-1}$

   e. $e^{-c} = 5^{2c}$ f. $e^{-t} = 0.04$ g. $\log_2(10 + 3x) = 5$ h. $\log x = 1$

   i. $\ln x - \ln(x - 4) = \log 12$ j. $\log_4(x+3) + \log_4(x-3) = 2$

4. Solve for the unknown variable in the following polynomial equations and inequalities. If no solution exits, say so. (*Hint: use Theorem 4.5.2, the distributive property with zero on one side and find common denominators when needed)*

   a. $x^2 \le 1$ b. $1/(x-3)=0$ c. $x/(x-3)<0$ d. $4-x^2 \le 0$

   e. $4x^3-7x^2 \le 15x$ f. $x/(x-5) \le 2$ g. $(x+1)/(x-2) \ge 3$

   h. $-4/(2x+7)<1/2$ i. $\frac{x-4}{x+3}-\frac{x+2}{x-1} \le 0$ j. $\frac{x^2-x-2}{x^2+5x+6}<0$

5. Try to estimate the value(s) of x for which $y=0$ by graphing the function.

   a. $y=-(2x)+7$ b. $y=x^3-x$ c. $y=\frac{3}{4}x^2+2$ d. $y=\frac{8x-4}{3x+6}$

6. Write z as a function of x.

   a. $z=ay^3+b,\ y=x-1$ b. $z=27y^3,\ y=(1/3)\sqrt[3]{x}$

   c. $z=a\sin(by+c)+d,\ y=bx^2$

7. Write z as a function of t.

   a. $z=ay^3+b,\ y=x-1,\ x=3t^5+k$ b. $z=27y^3-8,\ y=(1/3)\sqrt[3]{x},\ x=8t$

   c. $z=a\sin(by+c)+d,\ y=bx^2,\ x=\ln t^2$

   d. $z=y^2+1,\ y=x^2+1,\ x=t^2+1$

Solutions to Exercises

Chapter 1.

1. ----------?
2. Pigeons can't fly.
3. *If* $\underbrace{\textit{you read PM and you eat brains}}_{p}, \underbrace{\textit{then}}_{\Rightarrow} \underbrace{\textit{you are a psycho} - \textit{nerd or a zombie}}_{q}$
   a. $p \Rightarrow q$ is the original statement, therefore, the contrapositive is $\sim q \Rightarrow \sim p$: "if you're neither a psycho-nerd nor a zombie, then you don't read PM or you don't eat brains"; the converse is $q \Rightarrow p$: "if you are a psycho-nerd or a zombie, then you read PM and you eat brains"; the inverse ($\sim p \Rightarrow \sim q$) is "if you neither read PM nor eat brains, then you're neither a psycho-nerd nor a zombie".
   b. Here $p \Rightarrow q$ is the now the inverse, therefore, the converse must be $\sim q \Rightarrow \sim p$: "if you're neither a psycho-nerd nor a zombie, then you don't read PM or you don't eat brains; the original must be $\sim p \Rightarrow \sim q$: "if you don't read PM or you don't eat brains, then you're neither a psycho-nerd nor a zombie"; and the contrapositive must be $q \Rightarrow p$: "if you're a psycho-nerd or a zombie, then you read PM and you eat brains".(Notice that the converse of the original is the contrapositive of the inverse, among other connections.)
   c. The original and the contrapositive have equal truth values, and the converse and the inverse have equal truth values.
   d. "You read PM and you eat brains if and only if you are a psycho-nerd or a zombie" (or the reverse of this).
4. The negation of the phrases are as follows:
   a. It doesn't rain, or it doesn't snow.
   b. The sun doesn't shine, and the moon doesn't glow (or "neither the sun shines nor the moon glows).
   c. The boy doesn't like apples and doesn't like carrots, or he doesn't drink lemonade.
   d. The girl doesn't like spaghetti or doesn't like meatballs, and she likes soda.
5. The contrapositives are as follows:
   a. If s and $\sim t$, then $\sim p$ or $\sim q$.
   b. If $\sim t$ or $\sim w$, then (p or q) or $\sim s$.
   c. If t or w, then ($\sim p$ and q) or ($\sim r$ or s).
   d. If $\sim w$ or ($\sim v$ and $\sim x$), then $\sim p$ and (q and (r or s)).
6. Let p represent the statement *it rains in Spain,* and let q represent the phrase *the plains are dry* and s the phrase *clouds have gathered,* then the given exercise statements can be written as follows:
   a. $p \Leftrightarrow q \wedge s$ b. $q \wedge s \Rightarrow p$ c. $\sim q \vee \sim s \Rightarrow \sim p$ d. $\sim p \Rightarrow \sim s$

7. You must conclude that the victim was killed with (or by) a Mongolian yak.
8. $K = \{I,VI,VII,XIV,XV,XVI,XXV,XXVI\}$, the union is just K; the intersection is the set $\{VI,VII,XXV,XXVI\}$.
9. Determine the results.
   a. $\{a,b,c,d,e,f,g\}$ b. $\{a,b,c,d,e,f\}$ c. $\{c,d,e,f,g\}$
   d. $B = \{b,d,f,g\}$ e. $\{d,g,e\}$ f. $\varnothing$
10. Determine the truth or falsehood of the exercise statements, if the exercise statement is ill formed, explain why.
   a. True b. True c. Ill formed, a set cannot equal one of its elements (normally). d. True e. True f. false
   g. Ill formed, a set cannot be an element of a set (usually).
11. The sets $\varnothing$, $\{a\}$, $\{b\}$, $\{c\}$, $\{a, b\}$, $\{a, c\}$, $\{b, c\}$, $\{a, b, c\}$ are all subsets of the three-element set $\{a, b, c\}$: i.e., there are a total of eight subsets. And $\varnothing$, $\{a\}$, $\{b\}$, $\{c\}$, $\{d\}$, $\{a, b\}$, $\{a, c\}$, $\{a, d\}$, $\{b, c\}$, $\{b, d\}$, $\{c, d\}$, $\{a, b, c\}$, $\{a, b, d\}$, $\{a, c, d\}$, $\{b, c, d\}$, $\{a, b, c, d\}$ are all subsets of the four-element set $\{a, b, c, d\}$ for a total of sixteen subsets. The others are left for you. The formula involves a power of the number 2.
12. *All birds have feathers* is a general statement about what all birds have in common; however, not all birds have blue feathers, only certain specific birds have blue feathers. Logical variables in algebra behave similarly, they relate to the general characteristics of numbers. Thus, if x is a number, then $x+1=9$ is false because this is a characteristic specific to the number 8; on the other hand if x is a number, then $x+12=12+x$ is true because every number shares this property.

Chapter 2

1. See Theorem 2.2.1 and replace "times" with "plus".
2. It would be analogous to the multiplication axiom, that is E5.
3. We will proceed by example. (E1) Reflexive: every number is equal to itself because they will have the same remainder when divided by 2. (E2) Symmetric: if say $8=72$, then $72=8$ because changing which side of equality these numbers are on will not change their remainders. (E3) Transitive: if $8=72$ and $72=104$ (which they do), then $8=104$ because they will have the same remainders (likewise for odd numbers, if $3=45$ and $45=19$, then $3=19$). (E4) Addition: if $8=72$, then $8+3=72+3$. (I.e., no matter what number is added to equals, the sums will have the same remainder.) The same is true for multiplication. Axiom E6 also holds.
4. (a) Yes, all of the axioms of equality would hold. (b) The set of positive integers would be partitioned into three parts: the part with numbers whose remainders would be zero, the part with numbers whose remainders would be one, and the part with numbers whose remainders

would be two. (c) Yes, the equality axioms would still hold when the divisor is 4. (d) The positive integers would be partitioned into four sets.

5. You would not contradict the request if you picked out three marbles all the same color because they are *any* three marbles.
6. We need the symmetric axiom (E2) because without it we could not *logically* swap the positions of the numbers on either side of $=$. The equality axioms define *precisely* what the symbol $=$ can (and cannot) do.
7. (a) $2 \leq 2$ is a true statement, however, $2<2$ is false; therefore $a \leq b \not\Rightarrow a<b$. (b) $2 \leq 3$ is a true statement, however, $2=3$ is false; therefore $a \leq b \not\Rightarrow a=b$. (c) If $2=2$, then most certainly the number 2 is either less than 2 or equal to 2 (it equals 2). (d) This problem is a little tricky. Although it is true that $5 \leq 7$ and that $5<7$, it fails to meet the definition of $\leq$, which is $a \leq b$ *if and only if* $a<b$ *or* $a=b$. (e) This sentence is true by the law of trichotomy. (f) False: omitted was the option where $5<7$, which the Law of Trichotomy demands.
8. Both $-7<0$ and $0<7$ are true, and it is also true that $(-7)(-7)=7 \cdot 7$; however, $-7 \neq 7$. Thus, Theorem 2.2.1 is false if $a<0$ and $0<b$.
9. Find the solution set.
   a. $(1, 2)$ b. $[15, \infty)$ c. $(-\infty, 7) \cup \{10\}$ d. $[3, 10)$ e. $\{2, 9\}$
10. See Appendix A: ☆4
11. See Appendix A: ☆5
12. Prove that if $a<b$ and $c<d$, then $a+c<b+d$.
    Proof: By Axiom O3 (and because c is any number) $a+c<b+c$; by that same axiom (and because b is any number) $b+c<b+d$. Finally, the transitive axiom of inequalities, O2, allows us to write the required result: $a+c<b+d$.

Chapter 3

1. The set created is closed because once the integers 10 through 19 are generated (themselves being generated from the integers 3, 5, 6, 8, and 9), every integer after 19 is some sum of the integers 10 through 19. For example, $49=10+10+10+19$, and $63= 10+10+10+10+10+13$.
2. One possible table is the following:

+	α	β	γ	δ
α	α	β	γ	δ
β	β	α	β	γ
γ	γ	δ	α	β
δ	δ	γ	δ	α

The set {α, β, γ, δ} is closed according to the table. The set is not commutative because δ = γ + β ≠ β + γ = β (it only takes one counter example to disprove a general statement). The set is not associative because γ = (β + β) + γ≠ β + (β + γ) = α. The set has a zero element, which is α. Finally, each element has an additive inverse, which happens to be itself. If you cannot find a counter example in your table for a property, and you've exhausted every possibility, then the set abides by that property. [Question: do the elements of the table have to be symmetric (equal) on either side of the diagonal elements that run from the upper left to the lower right for the table to abide by the commutative axiom?]

Another possible table is the one using wrap-around addition (below), the details of which are left for you to explore.

+	α	β	γ	δ
α	α	β	γ	δ
β	β	γ	δ	α
γ	γ	δ	α	β
δ	δ	α	β	γ

3. Yes, the set $\{c\}$ is closed, and $c=0$.
4. Find the number of ways in which the expression $((\alpha+\beta)+\gamma)+\delta$ can be written using A2 and A3 pairwise. We have already shown that there are twelve ways to add three numbers, $\alpha, \beta, \gamma,$ pairwise using A2 and A3. We will now append, by adding, the number δ to the aft end and then to the fore end of each of these twelve ways; doing this will double our count to twenty-four. However, because we are working with four numbers, we could just as well have started with the three number α, β and δ (which would also have a count of twelve) and additively appended the number γ fore and aft of each of these, which would give us another twenty-four. Likewise, we could have started with the three numbers α, γ and δ and appended the number β fore and aft of each of these, giving us another twenty-four. Finally, we could have started with the numbers β, γ and δ and appended the number α for another twenty-four. When we add it all up, we get a grand total of four times twenty-four or ninety-six different ways that four numbers can be added. However, we need to be careful because we've missed a few: we've missed all sums of the type $(\alpha+\beta)+(\gamma+\delta)$ for each arrangement/sequence of the numbers $\alpha, \beta, \gamma, \delta$. Because there are twenty-four different arrangements of these four numbers, there are twenty-four different ways that we can write this type of sum. Thus, we must add another twenty-four to our count giving us a grand-grand total of

one hundred twenty ways to add four numbers.

5. $((\alpha+\beta)+\gamma)+\delta=\underbrace{(\alpha+(\beta+\gamma))}_{Associate}+\delta=(\underbrace{(\beta+\gamma)+\alpha}_{Commute})+\delta=((\underbrace{\gamma+\beta}_{Commute})+\alpha)+\delta=$

$\underbrace{(\gamma+\beta)+(\alpha+\delta)}_{Associate}=(\gamma+\beta)+(\underbrace{\delta+\alpha}_{Commute})$

6. It's all relative and symbols mean what we want them to mean.
7. It would have made no difference at all to mathematics if negative numbers had been chosen as being greater than positive numbers. Everything would work just fine because symbols only have the meanings we attach to them.
8. Simplify

   a. $x+2+a-b$ b. $y+x+6-b$ c. $x-z+45-a$ d. $x+y-13$
9. Solve for the logical variables x, y or z.

   a. $y=7-b$ b. $x=c+d+17$ c. $z=a+b+22$
10. Fill in the table using the equation $y=3x+9$. Some possible values are,

Independent variable	x	0	1	5	11
Dependent variable	y	9	12	24	42

11. All of the systems $[D,+]_A$, $[E_0,+]$, $[N_0,+]$ and $[C,\oplus]$ are Abelian groups. The system $[N,♣]$ is not a group because it doesn't abide by the associative axiom.
12. The set generated under addition from the elements of the set {0, 3, 5} is not a group, it lacks additive inverses. The set generated from the elements of {0, 5, −5} *is* an Abelian group.
13. All of the following generated sets are groups and subsets of C (the set of elements on a clock). The generated sets in problems c, e, and g are equal and equal the set C (the set of elements on a clock). Also, the generated sets in problems b and f are equal. Sets in problems a, b, d, and f only repeat the numbers shown.

    a. {0, 2, 4, 6, 8, 10} b. {0, 3, 6, 9}

    c. {0, 1, 2, 3, 4, 5, 6, 7, 8, 9, 10, 11, 12} d. {0, 6}

    e. {0, 1, 2, 3, 4, 5, 6, 7, 8, 9, 10, 11, 12} f. {0, 3, 6, 9}

    g. {0, 1, 2, 3, 4, 5, 6, 7, 8, 9, 10, 11, 12}
14. The generated sets are, {1}, {2}, {4}, {7}, {8}, {11}, {13}, {14}.

Chapter 4

1. a. closed b. not closed: the number 1 is not in the set. c. closed d. closed; it is the set of all odd integers because an odd number times an odd is odd; it is not a group because there are no multiplicative inverses in the set. e. {...,−8,−4,−2,0,2,4,8,...}; the set is not a group because there is neither a 1 nor multiplicative inverses in the set.

2.

a. To show that the set is indeed closed under both operations create the intersection and union tables, thus we have (respectively),

$\cap$	$\varnothing$	$\{a\}$	$\{a,b\}$	$\{a,b,c\}$
$\varnothing$	$\varnothing$	$\varnothing$	$\varnothing$	$\varnothing$
$\{a\}$	$\varnothing$	$\{a\}$	$\{a\}$	$\{a\}$
$\{a,b\}$	$\varnothing$	$\{a\}$	$\{a,b\}$	$\{a,b\}$
$\{a,b,c\}$	$\varnothing$	$\{a\}$	$\{a,b\}$	$\{a,b,c\}$

$\cup$	$\varnothing$	$\{a\}$	$\{a,b\}$	$\{a,b,c\}$
$\varnothing$	$\varnothing$	$\{a\}$	$\{a,b\}$	$\{a,b,c\}$
$\{a\}$	$\{a\}$	$\{a\}$	$\{a,b\}$	$\{a,b,c\}$
$\{a,b\}$	$\{a,b\}$	$\{a,b\}$	$\{a,b\}$	$\{a,b,c\}$
$\{a,b,c\}$	$\{a,b,c\}$	$\{a,b,c\}$	$\{a,b,c\}$	$\{a,b,c\}$

b. Every set contains the "nothing-element", but unfortunately there is no good way to exhibit "nothing". For example, the set $\{0\}$ contains the element 0, which we would write as $0 \in \{0\}$, it also contains the nothing-element, which, if one wanted to, could be written as " $\in \{0\}$" or perhaps as " $\_ \in \{0\}$ ". Of course, either way of expressing this is terribly awkward, so it is just not done. However, when it comes to subsets, not only is $\{0\} \subseteq \{0\}$, but also $\{\ \} \subseteq \{0\}$, or if you prefer, $\varnothing \subseteq \{0\}$. In the case of set S, however, not only does S contain the nothing element (every set does), it also contains the *set* that contains the nothing-element (which is just another way of saying the null set). Therefore, $\varnothing \subseteq S$ and $\varnothing \in S$.

c. No, the $\cap$-table lacks a unique identity element, while the $\cup$-table lacks inverses.

3. Generate the set, using multiplication. In answer to (2), every set is a subset of C; to fulfill the requirements of (1) a set must contain the *multiplicative* inverse of every number in the set (i.e., where the product of the two numbers equals 1). Because every set which has been multiplicatively generated from an even number will only produce even numbers, none of the products of any of the elements from these generated sets can ever have a remainder 1 when divided by 12. Therefore, the generated sets of problems a, d, and g, cannot be groups—they have no inverses.

a. $\{1,2,4,8\}$, not a group b. $\{1,3,9\}$, not a group c. $\{1,5\}$ this generated set is a group: 5 is its own multiplicative inverse. d. $\{0,1,6\}$, not a group. e. $\{1,7\}$ is a group, because $7\cdot 7$ (which equals 49) when divided by 12 has a reminder of 1, (i.e., $7\otimes 7=1$). f. $\{1,9\}$ is not a group. g. $\{1,4,10\}$ is not a group h. $\{1,11\}$ is a group.

4. The set generated from the number 7 using wrap-around multiplication with 13 is the set $\{1,2,3,4,5,6,7,8,9,10,11,12\}$, 0 is not in the set (why?). Because this set is a group, every number in the set is the multiplicative inverse of some other number in the set; for example, $5\otimes 8=1$, $9\otimes 3=1$ and $7\otimes 2=1$. (An arithmetic much like wrap-around, called *modular arithmetic*, was explored extensively by Carl F. Gauss.)
5. $2+3=(1+1)+(1+1+1)=1+1+1+1+1=5$ (by definition, substitution and associativity). Likewise, $2-3=2+(-3)=(1+1)+((-1)+(-1)+(-1))=$ $[1+(1+(-1))+(-1)]+(-1)=[1+(0)+(-1)]+(-1)=\text{etc.}=-1$.
6. Show or explain why.
   a. Because $y-2=-(2-y)=(-1)(2-y)$, we have $(-1)((2-y)/(2-y))=$ $(-1)(1)$, which yields the desired result.
   b. $\dfrac{\frac{a}{b}}{\frac{c}{d}}=\dfrac{a}{b}\cdot\dfrac{1}{\frac{c}{d}}=\dfrac{a}{b}\cdot\dfrac{1\cdot 1}{c\cdot\frac{1}{d}}=\dfrac{a}{b}\cdot\dfrac{1}{c}\cdot\dfrac{1}{\frac{1}{d}}=\dfrac{a}{b}\cdot\dfrac{1}{c}\cdot\dfrac{d}{1}=\dfrac{ad}{bc}$ (*Always search for recursive patterns*)
   c. $1\cdot 0=0$ is true either by axiom or by theorem, your pick, but axioms are always presumed to be the "stronger" of the two.
7. We have already shown by Theorem 4.6.3 that $\dfrac{a}{b}+\dfrac{c}{d}=\dfrac{ad+bc}{bd}$, but the numbers in the numerator and denominator are just numbers (*single* numbers), thus, $\dfrac{ad+bc}{bd}+\dfrac{e}{f}=\dfrac{[ad+bc]f+[bd]e}{[bd]f}$ using the same theorem.
8. Simplify; thus, when adding, create fractions that have a common denominator, or just use the fraction-addition theorem, which always works.
   a. $\dfrac{2}{3}+\dfrac{7}{6}+\dfrac{11}{12}+\dfrac{13}{24}=\dfrac{2}{3}\cdot\dfrac{8}{8}+\dfrac{7}{6}\cdot\dfrac{4}{4}+\dfrac{11}{12}\cdot\dfrac{2}{2}+\dfrac{13}{24}=\dfrac{16}{24}+\dfrac{28}{24}+\dfrac{22}{24}+\dfrac{13}{24}=\dfrac{79}{24}=$ $\dfrac{3\cdot 24+7}{24}=\dfrac{3\cdot 24}{24}+\dfrac{7}{24}=3\cdot\dfrac{24}{24}+\dfrac{7}{24}=3+\dfrac{7}{24}$, which, in short hand, is written as (i.e., defined as) $3\frac{7}{24}$.
   b. $\dfrac{3t+7t+5}{4t+16t+10}=\dfrac{10t+5}{20t+10}=\dfrac{5(2t+1)}{10(2t+1)}=\dfrac{1}{2}$, $(2t+1\neq 0)$
   c. $\dfrac{6y-3}{4y-2}+\dfrac{3z-9}{33-11z}=\dfrac{3(2y-1)}{2(2y-1)}+\dfrac{3(z-3)}{11(3-z)}=\dfrac{3}{2}+\dfrac{-3}{11}=\dfrac{33+(-6)}{22}=\dfrac{27}{22}=1\frac{5}{22}$
   d. $\dfrac{3a-2b}{2b}-\dfrac{3x-2y+7t}{7t}=\dfrac{3a-2b}{2b}+\dfrac{2y-3x-7t}{7t}=$

$$\frac{7t(3a-2b)+2b(2y-3x-7t)}{2b\cdot 7t}=\frac{21ta-14tb+4by-6bx-14bt}{14bt}=$$

$$\frac{(-2)14bt+21ta+4by-6bx}{14bt}=\frac{21at+4by-6bx}{14bt}-2,\ (b,t\neq 0)$$

e. $\frac{\frac{-ay}{x}+4y}{y}-\left(\frac{a}{-x}-3x\right)=\left(\frac{-ay}{x}+4y\right)\left(\frac{1}{y}\right)-\left(\frac{a}{-x}-3x\right)=$

$$\left(\frac{-a}{x}+4\right)+\frac{a}{x}+3x=3x+4,\ (x,y\neq 0)$$

f. $\frac{2n-k}{h}\cdot\frac{(2x+1)h}{x}\cdot\frac{z+1}{6n-3k}=\frac{\cancel{2n-k}^{1}}{\cancel{h}_{1}}\cdot\frac{(2x+1)\cancel{h}^{1}}{x}\cdot\frac{z+1}{3\,\cancel{(2n-k)}_{1}}=$

$$1\cdot\frac{2x+1}{x}\cdot\frac{z+1}{3}=\frac{2xz+2x+z+1}{3x},\ (h,\ x,\ 2n-k\neq 0)$$

g. $4(7h^{-1})^{-1}-\frac{h}{7}=4(7^{1(-1)}h^{(-1)(-1)})-\frac{h}{7}=4(7^{-1}h^{1})-\frac{h}{7}=4\left(\frac{1}{7}h\right)-\frac{h}{7}=$

$$\frac{4h}{7}-\frac{h}{7}=\frac{3h}{7},\ (h\neq 0)$$

h. $(x+\pi)(x-\pi)=x\cdot x\ \underbrace{\cancel{-x\cdot\pi}}_{0}\ \underbrace{\cancel{+\pi\cdot x}}_{0}-\pi\cdot\pi=x^2-\pi^2$

i. $(2x+3y+4)^{-1}(2x+3y-4)^{-1}=\frac{1}{2x+3y+4}\cdot\frac{1}{2x+3y-4}=$

$$\frac{1}{(2x+3y+4)(2x+3y-4)}=[(2x+3y+4)(2x+3y-4)]^{-1}=$$

$$[4xx+6xy\ \underbrace{\cancel{-8x}}_{0}+6yx+9yy\ \underbrace{\cancel{-12y}}_{0}\ \underbrace{\cancel{+8x}}_{0}\ \underbrace{\cancel{+12y}}_{0}-16]^{-1}=$$

$[4x^2+9y^2+12xy-16]^{-1}$, (naturally, the denominators do not equal zero). Like moves on a chess board, the choices one can make to solve algebra problems are vast; with practice, however, finding an expedient solution becomes second nature.

j. $(y-2x)^{-1}/[-7\cdot(14x-7y)^{-1}]=\frac{\frac{1}{y-2x}}{-7\cdot\frac{1}{14x-7y}}=\frac{1}{y-2x}\cdot\frac{1}{-7}\cdot\frac{1}{\frac{1}{14x-7y}}=$

$(14x-7y)/[-7(y-2x)]=\ [7(2x-y)]/[7(2x-y)]=1$. Another solution is $\frac{\frac{1}{y-2x}}{-7\cdot\frac{1}{14x-7y}}=\frac{\frac{1}{y-2x}}{\frac{7}{-(14x-7y)}}=\frac{\frac{1}{y-2x}}{\frac{7}{7(y-2x)}}=\frac{\left(\frac{1}{y-2x}\right)}{\left(\frac{1}{y-2x}\right)}=1,\ (y-2x\neq 0)$. If they look the same, they are the same.

9. --------?

Chapter 5

1. See Appendix A: ☆12
2. Perform the indicated operation.
   a. $\sqrt[7]{16j^3}\cdot\sqrt[7]{8j^6}\cdot\sqrt[7]{j}=\sqrt[7]{16\cdot 8\cdot j^3\cdot j^6\cdot j}=\sqrt[7]{2^7j^7j^3}=\sqrt[7]{2^7j^7}\cdot\sqrt[7]{j^3}=2j\cdot\sqrt[7]{j^3}$
   b. $(3-\sqrt{6x^3})(\sqrt{6x^3}+3)=3\sqrt{6x^3}+9-6x^3-3\sqrt{6x^3}=-6x^3+9$, $(x\geq 0)$
   c. $2\sqrt[3]{x^2}+\sqrt[3]{x^2}=3\sqrt[3]{x^2}$
   d. $\sqrt[3]{x-1}\cdot\sqrt[3]{1-x}\cdot\sqrt[3]{x-1}=\sqrt[3]{(x-1)(1-x)(x-1)}=\sqrt[3]{-(x-1)^3}=1-x$
   e. $5x\sqrt[3]{4xy}+3\sqrt[3]{2x^2}\cdot\sqrt[3]{2x^2y}=5x\sqrt[3]{4xy}+3\sqrt[3]{(2x^2)(2x^2y)}=$
   $5x\sqrt[3]{4xy}+3x\sqrt[3]{4xy}=8x\sqrt[3]{4xy}$
   f. $\sqrt[5]{-486x^2y^6}=-3y\sqrt[5]{2x^2y}$
3. Perform the multiplication.
   a. $\sqrt{3x}\cdot\sqrt[3]{4y^2}=(3x)^{1/2}(4y^2)^{1/3}=(3x)^{3/6}(4y^2)^{2/6}=[(3x)^3(4y^2)^2]^{1/6}=$
   $[27x^3\cdot 16y^4]^{1/6}=\sqrt[6]{432x^3y^4}$, $(x\geq 0)$
   b. $\sqrt{k}\cdot\sqrt[3]{k^2}\cdot\sqrt[6]{k^5}=\sqrt[6]{k^3}\cdot\sqrt[6]{k^4}\cdot\sqrt[6]{k^5}=\sqrt[6]{k^{12}}=k^2$, $(k\geq 0)$
   c. $\sqrt[5]{a^2b^3y}\cdot\sqrt[3]{ab^2x}\cdot\sqrt[15]{a^4b^{11}y^{12}}=\sqrt[15]{a^6b^9y^3}\cdot\sqrt[15]{a^5b^{10}x^5}\cdot\sqrt[15]{a^4b^{11}y^{12}}$
   $=\sqrt[15]{a^{15}b^{30}y^{15}x^5}=ab^2y\sqrt[15]{x^5}=ab^2y\sqrt[3]{x}$
   d. $(jk)^{1/2}\cdot(jk^3)^{1/5}\cdot(j^2k)^{3/7}=(jk)^{35/70}\cdot(jk^3)^{14/70}\cdot(j^2k)^{30/70}$
   $=(j^{35}k^{35}\cdot j^{14}k^{42}\cdot j^{60}k^{30})^{1/70}=jk(j^{39}k^{37})^{1/70}=jk\sqrt[70]{j^{39}k^{37}}$, $(j\cdot k\geq 0)$
   e. $\sqrt{x^2-1}\cdot\sqrt{x^2+x}\cdot\sqrt{x}=\sqrt{(x^2-1)(x^2+x)x}=$
   $\sqrt{x^2x^2x+x^2xx-x^2x-xx}=\sqrt{x^5+x^4-x^3-x^2}$, (radicands ≥ 0)
   f. $\sqrt{\sqrt{x-5}}\cdot\sqrt{\sqrt{x+5}}=\sqrt{\sqrt{(x-5)(x+5)}}=\sqrt[4]{x^2-25}$, (radicands ≥ 0)
4. Suppose instead that a rational times an irrational is a rational, then we may write, $(a/b)\cdot l=c/d$, where $a, b, c, d\in Z$ $(a, b, d\neq 0)$ and l is irrational. However, this implies that $l=(c/d)(b/a)=(cb)/(da)$, which is a rational number. This is a contradiction because an irrational number does not, by definition, equal a rational number—the conclusion follows.
5. Try $b=0.75$, then $1+0.75=1.75$ and $1/0.75=1.33...$, which indicates 0.75 is too large. Try $b=0.6$, then $1+0.6=1.6$ and $1/0.6=1.66...$, which indicates 0.6 is too small—but certainly closer. Try $b=0.65$, etc.
6. Suppose $\sqrt{a+b}=\sqrt{a}+\sqrt{b}$, then, upon squaring* both sides of this equation, we have $a+b=a+2\sqrt{ab}+b$. This, however, implies that $0=2\sqrt{ab}$, which implies that $0=\sqrt{ab}$. Squaring again implies that $0=ab$, which, by Theorem 4.5.2, implies either a or b must be equal to 0; therefore, if

---

* Technically speaking, we are not really "squaring", rather we are using Theorem 2.2.1 to multiply both sides by equals. We are multiplying the left side by $\sqrt{a+b}$ and the right side by $\sqrt{a}+\sqrt{b}$.

$\sqrt{a+b}=\sqrt{a}+\sqrt{b}$, then either $a=0$ or $b=0$ (or both). On the other hand, if either $a=0$ or $b=0$, say $b=0$, then clearly, $\sqrt{a+0}=\sqrt{a}+\sqrt{0}$ is a true statement. Thus, we have proved the *if and only if* statement.

7. a. $y\sqrt[3]{xy^2}$ b. $\dfrac{y^3a^2}{3}$ c. $b\sqrt[12]{b^5}+b^2\sqrt[12]{b}$ d. $\dfrac{|y^3|}{3\sqrt{3/a^4+2/b^4}}$

8. $2^{5^{(3^2)}}=2^{5^9}=2^{1953125}$, whereas $2^{(5^3)^2}=2^{(125)^2}=2^{15625}$. As you can see, these two numbers will not (do not) equal each other. This latter sequence of equalities is incorrect because we are taking the square of the cube of five, but that is not what we were originally specified to carry out. We were originally specified to take five to the power of the square of three.

9. a. 65536 b. $9y^3$ c. $(2z)^{16^{(1/2)^3}}=(2z)^{16^{(1/8)}}=(2z)^{\sqrt[8]{16}}=(2z)^{\sqrt{2}}$

Chapter 6

1. a. $\log_7 49=y$ b. $\log_{1/2} x=3$ c. $\log 100=4x$ d. $9^{2y}=81$
   e. $\pi^y=x$ f. $10^y=\log_\pi x$ g. $\ln y=\ln(2x+3)$; thus, $y=2x+3$.
2. i, ii, and iii (where applicable) are equal for each row.
   a. ~ 5.6 b. 2 c. 0 d. 0 e. ~ 3.4 f. ~ 1.2 g. $\sim -.65$
3. a. $3x-7=49 \Rightarrow x=18\tfrac{2}{3}$ b. $2^{4x-6}=2^6 \Rightarrow x=3$ c. $3^3=3^{5x}\cdot(3^2)^{x^2} \Rightarrow$
   $3=5x+2x^2 \Rightarrow 2x^2+5x-3=0 \Rightarrow (2x-1)(x+3)=0 \Rightarrow x=1/2$ or $x=-3$
   d. $x\log 3=(x-1)\log 2 \Rightarrow \dfrac{x}{x-1}=\dfrac{\log 2}{\log 3}\approx 0.631 \Rightarrow x=-1.71$
   e. $-c=2c\ln 5$ (be careful, if $c=0$ you cannot divide. In fact, the only number which will make the equation true is $c=0$. (Why?)
   f. $-t=\ln(0.04) \Rightarrow t\approx 3.22$ g. $2^5=32=10+3x \Rightarrow x=22/3$ h. $x=10$
   i. $x/(x-4)\approx e^{1.08} \Rightarrow x\approx 6.06$ j. $(x+3)(x-3)=4^2 \Rightarrow x=\pm 5$
4. a. $x\in[-1,1]$ b. False (no solution) c. $0<x<3$ d. $x\notin(-2,2)$
   e. $0\le x\le 3$ or $x\le -5/4$ f. $x\in(-\infty,5)\cup[10,\infty)$ g. $x\in(2,7/2]$
   h. $x\notin[-15/2,-7/2]$
   i. $\dfrac{(x-4)(x-1)-(x+2)(x+3)}{(x+3)(x-1)}\le 0$, simplify this inequality and use logic about the factors. Answer: $-3<x\le -1/5$ or $x>1$.
   j. $x\in(-3,-2)\cup(-1,2)$
5. a. $x=7/2$ b. $x=0, 1$ c. False (no solution for $y=0$) d. $x=1/2$
6. a. $z=a(x-1)^3+b$ b. $z=x$; here y is the inverse function of z. It undoes what the function z did.
   c. $z=a\sin(b^2x^2+c)+d$
7. a. $z=a((3t^5+k)-1)^3+b$ b. $z=8t-8$ c. $z=a\sin(b^2\ln^2 t^2+c)+d$; (*note*: $\ln^2 x\equiv(\ln x)^2$) d. $z=((t^2+1)^2+1)^2+1$

APPENDIX A
ANSWERS TO STAR (☆) QUESTIONS

☆1. Write the converse and the inverse of,
"If you read PM and you eat brains, then you are a zombie".
Converse: If you are a zombie, then you read PM and you eat brains.
Inverse: If you don't read PM or you don't eat brains, then you are not a zombie.

☆2. Show that $((A\cap B)\cup(B\cap C))\cap(A\cap B\cap C)'=\{d,f\}$.
$A\cap B=\{d,g\}$, $B\cap C=\{f,g\}$; thus, their union is the set $\{d,f,g\}$. Also $(A\cap B\cap C)=\{g\}$, and its complement is $(A\cap B\cap C)'=\{a,b,c,d,e,f\}$. Thus, $\{d,f,g\}\cap\{a,b,c,d,e,f\}=\{d,f\}$.

☆3. Prove that if $aaaa=bbbb$ then $a=b$. The product of four factors equals the product two factors composed of two factors each, hence $(aa)(aa)=(bb)(bb) \Rightarrow aa=bb$ and by theorem 2.2.1 $a=b$.

Prove that if $aaaaaa=bbbbbb$ (six factors), then $a=b$. We have already proved the theorem for three factors and for two factors and the product of six factors equals the product of three factors composed of two factors each (or two composed of three each), thus $a=b$.

The product of eight factors equals the product of two factors, each of which is composed of two factors, and where each of these is composed of two factors, thus, $((aa)(aa))((aa)(aa))=((bb)(bb))((bb)(bb))\Rightarrow a=b$.

☆4. Suppose $a<0$ and $b<0$, then $0<-a$ and $0<-b$; now suppose that $(-a)(-a)(-a)=(-b)(-b)(-b)$, but that $-a\neq -b$, then, following the example from page 45, we would discover that this leads us into a contradiction. Thus, we must conclude that $-a=-b$, but this implies that $a=b$. Furthermore, $(-a)(-a)(-a)=(-b)(-b)(-b)$ implies $(a)(a)(a)=(b)(b)(b)$. The converse is somewhat trivial and is left for you to verify. Thus, given the numbers a and b, $a<0$, $b<0$, then $a\cdot a\cdot a=b\cdot b\cdot b$ if and only if $a=b$.

☆5. Because without the condition of $0<c$ (i.e., if c were any number) for Axiom O4, our theorems would culminate in a contradiction. For example, suppose we have $-3<-2$, then upon adding 2 to both sides followed by adding 3 to both sides (Axiom O3), we would have $2<3$; however, multiplying both sides of $-3<-2$ by -1, which would be legal in the absence of the condition $0<c$, we would have $3<2$. The Law of Trichotomy states that we cannot have both. Thus, without the condition we have a contradiction.

☆6. You will be shown a pattern for finding the answer to the number of ways ten numbers can be added, then you must follow that pattern to ten numbers on your own.

There are only two ways to add two numbers: $a+b$ or $b+a$ and twelve ways to add three numbers. Thus, for four numbers, one needs to append (add) the new, fourth number to the left or to the right of the first three numbers: $(a+b+c)+d$ or $d+(a+b+c)$. That would double the count of just three numbers to 24. However, *any* of the four numbers could have been on the "outside" of the other three. Thus, we would have $4\times 24=96$ different ways. However, in our analysis, we have neglected the sum which includes a pair of pairs. In other words, we have neglected all sums of the sort $(a+b)+(c+d)$ for every permutation of the four numbers, which works out to 24 more sums. Thus, the total number of ways that four numbers can be added is 120. The number of ways that five numbers can be added can be found by appending the new number to the left or right of the four numbers: $(a+b+c+d)+e$ or $e+(a+b+c+d)$, etc. Find the pattern, and the rest is up to you, and don't forget to look for those configurations that were missed.

☆7. Prove that $-(-a+(-b)+(-c))=a+b+c$.
Proof: $-(-a+(-b)+(-c))=-((-a+(-b))+(-c))$. By Theorem 3.5.3, we have $-((-a+(-b))+(-c))=-(-a+(-b))+c$. Applying Theorem 3.5.3 again to the right-hand side of this equality, will yield the desired result.

☆8. Show that $Z\oplus Z^{-1}=Z\otimes Z^{-1}$. Every fraction, a/b, where a and b are integers ($b\neq 0$), can be written as $1/b+1/b+\cdots+1/b$, where the sum of all the 1s in the numerators equal the integer a. This sum, however, is an element of the set $Z\oplus Z^{-1}$ because $1/b$ is an element of that set (why?). However, the fraction, a/b, is also an element of the set $Z\otimes Z^{-1}$ because $a\cdot(1/b)\in Z\otimes Z^{-1}$. Thus, $a/b\in Z\oplus Z^{-1}\Rightarrow a/b\in Z\otimes Z^{-1}$. Consequently, $Z\oplus Z^{-1}\subseteq Z\otimes Z^{-1}$. Because every integer can be written as a fraction, i.e., $a/1$, this result is also true for every integer. I leave it to you to convince yourself that $Z\otimes Z^{-1}\subseteq Z\oplus Z^{-1}$. (*You will need the distributive axiom.*)

~~☆9~~. BLANK

☆10. Prove that $(a^m)^n=a^{mn}$. There are n factors of m factors of a. That is to say $(a^m)^n=\underbrace{a^m a^m a^m \cdots a^m}_{n\text{ times}}$, where each a^m represent m factors of a; thus, there are $n\times m$ factors of a (i.e., a^{mn}).

☆11. Prove that $\sqrt[m]{a}\cdot\sqrt[n]{b}=\sqrt[mn]{a^n b^m}$. A proof of this sort, which requires us to show that one side of an equality can be made to look identically like the other side, must be accomplished by applying the axioms, theorems and definitions of algebra to one side of the equality only. In this case, we will make the left side of the equality look like the right. With that in mind, if we multiply together mn factors of $\sqrt[m]{a}$ with mn factors of $\sqrt[n]{b}$, then we will have an equal number of factors of both roots. Thus, $(\sqrt[m]{a})^{mn} \cdot (\sqrt[n]{b})^{mn} = a^n b^m$. Also, $(\sqrt[m]{a})^{mn} \cdot (\sqrt[n]{b})^{mn} = (\sqrt[m]{a} \cdot \sqrt[n]{b})^{mn}$. Therefore, by the transitivity of equality axiom, we have $(\sqrt[m]{a} \cdot \sqrt[n]{b})^{mn} = a^n b^m$. However, the definition of roots is such that $(\sqrt[mn]{a^n b^m})^{mn} = a^n b^m$. Thus, $(\sqrt[m]{a} \cdot \sqrt[n]{b})^{mn} = (\sqrt[mn]{a^n b^m})^{mn}$ by transitivity of equals, and hence, $\sqrt[m]{a} \cdot \sqrt[n]{b} = \sqrt[mn]{a^n b^m}$ by Theorem 2.2.1 (Extended).

☆12. Prove that $\sqrt{3}$ is irrational. As was done with the $\sqrt{2}$, we may assume that $\sqrt{3} = a/b$, where a and b are integers with no factors in common, then, because $\sqrt{3} = a/b$, we may write (as was done with $\sqrt{2}$) $a \cdot a = 3 \cdot b \cdot b$. ($a \cdot a$ and $b \cdot b$ are integers, as is $3 \cdot b \cdot b$, because they are the products of integers.) Suppose, now, that $a \cdot a$ is even, then $3 \cdot b \cdot b$ must be even (an even number can only equal an even number), but that can only be the case if $b \cdot b$ is even, in which case a and b have the factor 2 in common. However, this contradicts our assumption that the fraction a/b was in lowest terms and had no common factors. Therefore, $a \cdot a$ and $b \cdot b$ must both be odd. Now, the equation $a \cdot a = 3 \cdot b \cdot b$ tells us that $a \cdot a$ is a number that contains the factor 3; in other words 3 divides into $a \cdot a$ without a remainder (i.e., $(a \cdot a)/3 = b \cdot b$, where $b \cdot b$ is an integer) or with a remainder 0 if you prefer. Before we continue, however, we must mention an important point: every number which is *not* divisible by 3 must yield a remainder of 1 or 2 when divided by 3. We will accept this fact as intuitively obvious.

We now show, by contradiction, that if $a \cdot a$ is divisible by 3, then a is divisible by 3. If $a = 0$ the result immediately follows. So, suppose that $a > 0$ is not divisible by 3, then either $a = h + 1$, where h is divisible by 3 or $a = k + 2$, where k is divisible by 3, for some h or some k. In each case the remainder when a is divided by 3 is either 1 or 2 (respectively). However, $a \cdot a = (h+1)(h+1) = hh + h + h + 1$ is a number that has a remainder of 1 when divided by 3 [because each h is divisible by 3 —recall that $(a+b)/c = a/c + b/c$], contrary to our premise that $a \cdot a$ has a remainder of 0 when divided by three. Likewise, $a \cdot a = (k+2)(k+2) = kk + 2k + 2k + 3 + 1$ has a remainder of 1 when divided by 3 (because the first four terms are divisible by 3), which also contradicts our premise that $a \cdot a$ has a remainder of 0 when divided by 3. Thus, because our assumption that a not divisible by 3 results in a contradiction, it must be the case that a *is* divisible by 3; in other words, a has a factor of 3, say $a = 3n$; so, we may write $a \cdot a = 3n \cdot 3n = 9 \cdot n \cdot n$, but recall that $a \cdot a = 3 \cdot b \cdot b$; therefore, $9 \cdot n \cdot n = 3 \cdot b \cdot b$, which implies $3 \cdot n \cdot n = b \cdot b$. By the same reasoning as before, we must conclude that b is also divisible by 3, which means that a and b each contain the common factor 3, which contradicts our assumption that a and b have no common factors. Therefore, $\sqrt{3}$ is not equal to a rational number. This method can be used to show that the square root of any prime number is irrational (try your hand with $\sqrt{5}$, and the remainders of 5). What about higher roots?

APPENDIX B

CONSTRUCTING THE SET $Z \otimes Z^{-1}$

Constructing the set $Z \otimes Z^{-1}$ can make for a very lively classroom discussion, by providing students with physical objects which have distinct characteristics, like coins, crayons, and colored bingo chips or what have you, and having the students record the various parings that can be made from two such sets (of course, this can just as easily be done with numbers on paper).

For this construction, I will present the set of integers to the left of the page and the set of multiplicative inverses of the integers to the right. Then the set $Z \otimes Z^{-1}$, will just be the product of all the (line) parings of the numbers from the left set to the right set as shown (or the right side to the left side if you prefer).

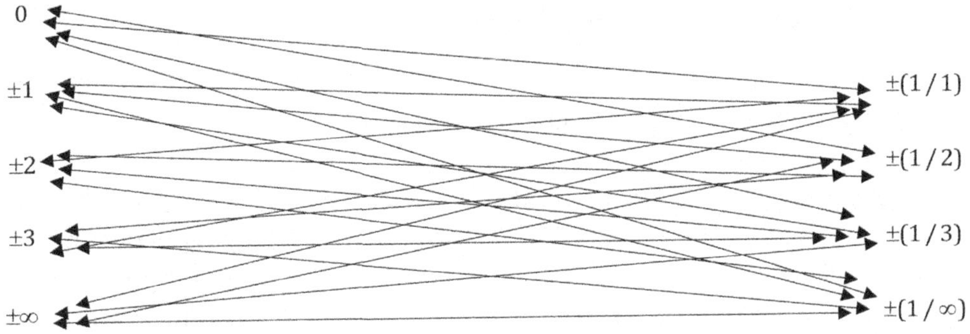

The set that corresponds to this construction is the following:

$$Z \otimes Z^{-1} = \{0,\ \pm 1 \cdot \frac{1}{1},\ \pm 1 \cdot \frac{1}{2},\ \cdots\ \pm 2 \cdot \frac{1}{1},\ \pm 2 \cdot \frac{1}{2},\ \cdots\ \pm 3 \cdot \frac{1}{1},\ \pm 3 \cdot \frac{1}{2}\ \cdots\ \pm \infty \cdot \frac{1}{\infty}\}$$

Although I have removed a lot of the repeats regarding the sign and the products of numbers which involve zero, the set still contains a lot of equivalent fraction repeats, for example $3 = 9 \times (1/3)$. Unfortunately, this makes it difficult to discuss this set of numbers from a logical perspective. Consequently, mathematicians have found other methods for constructing the set of rational numbers. For our purposes, however, we will just assume that all equivalent fractions have been removed from the set leaving only fractions that are in their most reduced form.

I will leave it for you to convince yourself that

$$Z \otimes Z^{-1} = (Z_P \otimes Z^{-1}) \cup \{0\} = (Z_N \otimes Z^{-1}) \cup \{0\}.$$

APPENDIX C

AXIOMS, THEOREMS, DEFINITIONS

(Tab this page for quick reference)

AXIOMS OF COMPARISON: CHAPTER 2

❖ THE AXIOMS OF EQUALITY

For any numbers a, b, *and* c

1.	$a = a$	Reflexive Property
2.	*If* $a = b$, *then* $b = a$	Symmetric Property
3.	*If* $a = b$ *and* $b = c$, *then* $a = c$	Transitive Property
4.	*If* $a = b$, *then* $a + c = b + c$	Addition Property
5.	*If* $a = b$, *then* $a \cdot c = b \cdot c$	Multiplication Property
6.	*If* $a = b$, *then* a *may replace* b *(or* b *may replace* a *) in any expression, formula or statement that contains* b *(or ... that contains* a*)*	Substitution Property

❖ THE AXIOMS OF INEQUALITY (ORDER AXIOMS)

For any numbers a, b, *and* c

1.	*Exactly one of the following conditions must hold for any two number:* $a = b, a < b$ or $b < a$	Law of Trichotomy
2.	*If* $a < b$ *and* $b < c$, *then* $a < c$	Transitive Property
3.	*If* $a < b$, *then* $a + c < b + c$	Addition Property
4.	*If* $a < b$ *and* $0 < c$, *then* $a \cdot c < b \cdot c$	Multiplication Property

And we define "$>$", which is read as *"is greater than"*, as $a > b$ if and only if $b < a$.

◈ THEOREM 2.2.1

If a *and* b *are numbers,* $0 < a$ *and* $0 < b$, *and if* $a \cdot a = b \cdot b$, *then* $a = b$.

◈ THEOREM 2.2.1, EXTENSION

If a *and* b *are numbers, where* $0 < a$ *and* $0 < b$, *and if* $a \cdot a \cdots a = b \cdot b \cdots b$, *where the number of* a *factors equals the number of* b *factors, then* $a = b$.

ADDITION AXIOMS: CHAPTER 3

∘ DEF. 3.1.1: SET CLOSURE

A set is closed under an operation of mathematics, if whenever we apply the operation to any two elements from within the set, the result is another element in the set.

❖ AXIOM A1: CLOSURE UNDER ADDITION

If a *and* b *are numbers, then* $a + b$ *is a number*

❖ **AXIOM A2: COMMUTATIVITY OF ADDITION**
If a and b are numbers, then $a+b=b+a$

❖ **AXIOM A3: ASSOCIATIVITY OF ADDITION**
If a, b and c are numbers, then $a+(b+c)=(a+b)+c$

❖ **AXIOM A4: THE EXISTENCE OF THE ADDITIVE IDENTITY NUMBER**
There exists a number called zero, which is written as "0", such that if a is a number, then $a+0=a$

❖ **AXIOM A5: THE EXISTENCE OF ADDITIVE INVERSES**
For every number, a, there is a number called the additive inverse of a, written as $-a$, such that $a+(-a)=0$.

◈ **THEOREM 3.5.1**
$0=-0$

◈ **THEOREM 3.5.2**
$-(-a)=a$

◈ **THEOREM 3.5.3**
$-(a+b)=-a+(-b)$

◇ **COROLLARY 3.5.3**
$-(a+b+c)=-a+(-b)+(-c)$, *and* $-(-a+(-b)+(-c))=a+b+c$.

∘ DEF. 3.5.1: SUBTRACTION
For numbers a and b, we define $a-b=a+(-b)$

MULTIPLICATION AND DISTRIBUTION AXIOMS: CHAPTER 4

❖ **AXIOM M1: CLOSURE UNDER MULTIPLICATION**
If a and b are numbers, then ab is a number

❖ **AXIOM M2: COMMUTATIVITY OF MULTIPLICATION**
If a and b are numbers, then $ab=ba$

∘ DEF. 4.2.1: ORDER OF OPERATIONS: MULTIPLICATION OVER ADDITION
The commutative property of multiplication shall take precedence over the commutative property of addition in the absence of parentheses: for numbers a, b, and c, we define $a+b\cdot c=a+(b\cdot c)$ and $a\cdot b+c=(a\cdot b)+c$

❖ **AXIOM M3: ASSOCIATIVITY OF MULTIPLICATION**
If a, b and c are numbers, then $a(bc)=(ab)c$

❖ **AXIOM M4: THE EXISTENCE OF THE MULTIPLICATIVE IDENTITY NUMBER**
There exists a number called one, which is written as "1", such that if a is a number, then $a\cdot 1=a$

◦ DEF. 4.4.1: THE SET OF POSITIVE INTEGERS

$Z_P = \{ 1,\ 1+1=2,\ 1+1+1=3,\ 1+1+1+1=4,\ 1+1+1+1+1=5,\ \dots \}$

◦ DEF. 4.4.2: THE SET OF NEGATIVE INTEGERS

$Z_N = \{ \dots , (-1)+(-1)+(-1)=-3, (-1)+(-1)=-2,\ -1\}$

◦ DEF. 4.4.3: THE SET OF INTEGERS

$Z = Z_P \cup Z_N \cup \{0\} = \{ \dots -3,\ -2,\ -1,\ 0,\ 1,\ 2,\ 3,\ \dots \}$

◆ **THEOREM: SUM OF INTEGERS**

The sum of two integers is an integer.

❖ **AXIOM M5: THE EXISTENCE OF MULTIPLICATIVE INVERSES**

For every number a, $a \neq 0$, there is a number called the multiplicative inverse of a, written as $\frac{1}{a}$ or a^{-1} (or $1/a$) such that $a \cdot \frac{1}{a} = 1$ or $a \cdot a^{-1} = 1$

◆ **THEOREM 4.5.1**

$$\frac{1}{\left(\frac{1}{a}\right)} = (a^{-1})^{-1} = a,\ a \neq 0$$

◆ **THEOREM 4.5.2 AND ITS COROLLARY**

If $a \neq 0$ and $b \neq 0$ are numbers, then $ab \neq 0$

If $ab = 0$, then either $a = 0$ or $b = 0$, or both.

◆ **THEOREM 4.5.3**

For numbers $a \neq 0$ and $b \neq 0$, $\frac{1}{a} \cdot \frac{1}{b} = \frac{1}{a \cdot b}$

◦ DEF. 4.5.1: DIVISION AND QUOTIENTS

For numbers a and b $(b \neq 0)$, we define the quotient of a and (or with, or divided by) b as $\frac{a}{b} = a \cdot \frac{1}{b}$

◦ DEF. 4.5.2: RATIONAL NUMBERS

A rational number is any number that can be expressed as the quotient of two integers (0 not in the denominator). The set of all rational numbers is represented with the symbol Q.

◆ **THEOREM 4.5.4**

For numbers a, and $b \neq 0$, $-\frac{a}{b} = \frac{-a}{b} = \frac{a}{-b}$

❖ **AXIOM D: DISTRIBUTION**

For numbers a, b and c, $c(a+b) = ca + cb$ *and* $(a+b)c = ac + bc$

◆ **THEOREM 4.6.1**

For every number a, $a \cdot 0 = 0$

◆ **THEOREM 4.6.2**

For every number a, $(-1) \cdot a = -a$,

◆ **THEOREM: PRODUCT OF INTEGERS**

The product of two integers is an integer; and, by extension, the product of two multiplicative inverses of integers is a multiplicative inverse of an integer

◆ **THEOREM 4.6.3**

For any numbers a, b, c *and* d, *where* $b \neq 0$ *and* $d \neq 0$, $\frac{a}{b} + \frac{c}{d} = \frac{ad+bc}{bd}$

◆ **THEOREM 4.6.4**

If $0 < a$, *then* $0 < \frac{1}{a}$; *and if* $a < 0$, *then* $\frac{1}{a} < 0$

<u>IRRATIONAL AND REAL NUMBERS, ROOTS AND EXPONENTS: CHAPTER 5</u>

∘ DEF. 5.1.1: IRRATIONAL NUMBERS

An irrational number is any number that <u>cannot</u> be expressed as the quotient of two integers.

∘ DEF. 5.1.2: THE SQUARE ROOT OF A NUMBER

The square roots of a number, $a \geq 0$, *written as* $\pm\sqrt{a}$ *or* $\pm\sqrt[2]{a}$, *are the numbers that when multiplied by themselves are equal to the number* a. *Thus, because* $\sqrt{a} \cdot \sqrt{a} = a$ *and* $(-\sqrt{a}) \cdot (-\sqrt{a}) = a$, $\sqrt{a}$ *and* $-\sqrt{a}$ *are square roots of* a. $\sqrt{a} \geq 0$ *is called the principal square root of* a.

◆ **THEOREM 5.1.1**

If $0 \leq a$ *and* $0 \leq b$, *then* $\sqrt{a} \cdot \sqrt{b} = \sqrt{ab}$

∘ DEF. 5.1.3: THE nth ROOT OF A NUMBER

The nth *root of a number,* a ($n \in Z_P$, $n \geq 2$, *and* $a \geq 0$ *if* n *is even), written as* $\sqrt[n]{a}$ *is the number that when multiplied by itself* n *times is the number* a. *Thus,*

$$\underbrace{\sqrt[n]{a} \cdot \sqrt[n]{a} \cdot \sqrt[n]{a} \;\ldots\; \cdot \sqrt[n]{a}}_{n \text{ factors of } \sqrt[n]{a}} = a.$$

◆ **THEOREM 5.1.2**

$\sqrt[n]{a} \cdot \sqrt[n]{b} = \sqrt[n]{ab}$ ($n \in Z_P$, *and* $n \geq 2$), *provided* $a \geq 0$ *and* $b \geq 0$ *whenever* n *is even.*

∘ DEF. 5.2.1: EXPONENTS

The nth *power of a number* ($n \in Z_P$, *and* $n \geq 1$), a, *written as* a^n, *is the number* a *multiplied by itself* n *times. Thus,*

$$\underbrace{a \cdot a \cdot a \cdot \;\ldots\; \cdot a}_{n \text{ factors of } a} = a^n.$$

The number n is called the exponent and the number a is called the base and $a^1 = a$.

◆ **THEOREM 5.2.1**

$(a^m)^n = a^{m \cdot n}$.

◆ **THEOREM 5.2.2**

$$a^{m^n} = a^{\overbrace{m \cdot m \cdot m \cdots m}^{n \text{ factors of } m}} \quad \text{(Note that } a^{m^n} \neq (a^m)^n\text{)}$$

∘ DEF. 5.2.2: EXPONENTS AS ROOTS

$$a^{1/n} = \sqrt[n]{a}.$$

∘ DEF. 5.2.3: DEFINITION OF EXPONENTS EXTENDED

The j/k power of a number a, written as $a^{j/k}$, (where $j/k \in Q$ is in reduced form and $a \geq 0$ whenever k is even), is the kth root of the number a multiplied by itself j times. Thus, $a^{j/k} = (\sqrt[k]{a})^j$ or alternatively $a^{j/k} = \sqrt[k]{a^j}$.

◆ **THEOREM 5.2.3**

$\sqrt[m]{a} \cdot \sqrt[n]{b} = \sqrt[(mn)]{a^n b^m}$ *or equivalently* $a^{\frac{1}{m}} \cdot b^{\frac{1}{n}} = (a^n b^m)^{\frac{1}{mn}}$.

CONTINUED STUDY: CHAPTER 6

∘ DEF. 6.1.1: LOGARITHM

For $x, b > 0$, $\log_b x = y$ if and only if $b^y = x$.

◆ **THEOREM 6.1.1**

Rule No. 1: $\log_b(ac) = \log_b a + \log_b c$

◆ **THEOREM 6.1.3**

Rule No. 3: $\log_b a^{-x} = -x \log_b a$ (This follows directly from Theorem 6.1.2.)

◆ **THEOREM 6.1.2**

Rule No. 2: $\log_b a^x = x \log_b a$

∘ DEF. 6.4.1: FUNCTION

A function from set A to set B is a correspondence that relates each element of set A to exactly one element of set B.

∘ DEF. 6.5.1: IMAGINARY NUMBER RULES/DEFINITIONS

If $0 < a, b$, and n is an even, positive integer, then $\sqrt{-a} = i\sqrt{a}$ and $\sqrt[n]{(-a)(-b)} \neq \sqrt[n]{-a} \cdot \sqrt[n]{-b}$, where $i = \sqrt{-1}$; $i^0 = 1$; $i^1 = i$; $i^2 = -1$; $i^3 = -i$; and $i^4 = 1$ $(= i^0)$. Otherwise, imaginary numbers abide by every algebraic axiom and definition.*

---

* It is customary to put the symbol ***i*** before a radical and a variable, but after a numeral number.

Acknowledgements

I would like to thank Ginny Maury for doing a proof read of the first edition of this book, her stamp is therefore on this edition. An English-major graduate and retired elementary school teacher, she had graciously taken on the unenviable task of proof reading a book on mathematics. Kudos to you and thank you Ginny!

I would also like to give a sincere thank you out to Walter Orange, a retired professor of the University of Pittsburgh, who also graciously undertook the task of proof reading this book. His detailed corrections and feedback for a large section of this book are greatly appreciated. Professor Orange is also a math historian buff. Our discussions and inquiries into the history of mathematics over a glass of fine wine and one of his wife's marvelous baked delicacies has always been a treasured delight of mine.

However, the job of proof reading is never done, and the buck always stops with the author. Any errors or omissions rest solely with me (and may they rest in peace).

Finally, I would also like to thank my lovely wife, Jana, who during the many years I have spent working on this book never once complained. On the contrary, she brought me tea, toast and a kiss. Thank you, Girl. I love you.

Bibliography

Allenby, R. (1983). *Rings, Fields and Groups.* Great Britain: Arnold.

Aristotle. (2011). *The Philosophy of Aristotle* (Reissue ed.). (A. E. Creed, Trans.) Signet.

Brown, D. S., 2), A. S., Boston, Littel, M., & editions), (. (1997). *Algebra: Structure and Method (Books 1 & 2);Boston; ; (all Ed.)* (All Editions ed.). Boston: McDougal Littel.

Dowden, B. H. (2014). *Logical Reasoning.* Sacramento, CA: California State University.

Doyle, S. A. (1994). *The Hound of the Baskervilles* (Thrift ed.). Mineola: Dover.

Ernest Nagel. (2001). *Godel's Proof.* (J. R. Newman, Ed.) N.Y.: New York University Press.

Gulberg, J. (1997). *Mathematics from the Birth of Numbers.* New York: W.W. Norton & Company, Inc.

Harel, G. (2014, January). *Common Core State Standards for Geometry: An Alternative Approach.* Retrieved from Ams.org: http://www.ams.org/notices/201401/rnoti-p24.pdf

Hofstadter, D. (1979). *Godel, Escher, Bach: an Eternal Golden Braid.* New York: Basic Books.

Katz, V. J. (1993). *A History of Mathematics: An Introduction.* New York: Harper Collins College.

Kleiner, I. (1991). Rigor and Proof in Mathematics: A Historical Perspective. *Mathematics Magazine, 64*(5), pp. 291-314.

Mueller, I. (1969, December). Euclid's Elements and the Axiomatic Method. *The British Journal for the Philosophy of Science, 20*(4), 289-309. Retrieved 2017, from http:/www.jstor.org

Nagel, E., & Newman, J. R. (2001). *Goedel's Proof* (Rev. ed.). (D. R. Hofstadter, Ed.) New York: New York University Press.

Natalia Marmasse, A. B. (2000, May 11). *Media.Mit.edu.* Retrieved from Numerical Mechanisms and Children's Concept of Numbers: http://web.media.mit.edu/~stefanm/society/som_final.html

Nicholson, W. K. (1993). *Introduction to Abstract Algebra.* Boston: PWS Publishing Co.

Niven, I. (1961). *Numbers: Rational and Irrational .* New York: Random House, Inc.

Poincare, H. (1969, March). Intuition and Logic in Mathematics. (F. J. Crosswhite, Ed.) *The Mathematics Teacher, 62*(3), 205-212.

Ray C. Jurgensen, A. J. (1975). *Geometry.* Boston: Houghton Mifflin Company.

Schwartz, R. (2014, September 17). *Dedekind Cuts.* Retrieved from Math.Brown.Edu: https://www.math.brown.edu/~res/INF/handout3.pdf

Seife, C. (2000). *Zero: The Biography of a Dangerous Idea.* New York: Penguin Books.

Smullyan, R. M. (2014). *A Beginners's Guide to Mahtematical Logic.* Mineola: Dover Publications, Inc.

Sullivan, S. M. (2006). *An Introduction to Traditional Logic: Classical Reasoning for Contemporary Minds.* North Charleston: BookSurge.

Suppes, P. (1965). *The Axiomatic Method in High-School Mathematics.* Stanford California: Institute for Mathematical Studies in the Social Sciences; Stanfor University.

Tweddle, J. C. (2012, January 6). *Weierstrass's Construction of the Irrational Numbers.* Retrieved from Faculty.Evansville.edu: http://faculty.evansville.edu/ct55/Portfolio/WeierstrassArticle.pdf

Victor J. Datz, K. H. (2014). *Taming the Unknown: A History of Algebra from Antiquity to the Early Twentieth Century.* New Jersey: Princeton University Press.

White, D. (2015). Connections between Abstract Algebra and High School Algebra: A few Connections Worth Exploring. *American Mathematical Society Blogs.* Retrieved Dec 2017, from Blogs.AMS.org

Whitehead, A. N., & Bertrand, R. (1910, 1912, 1913). *Principia Mathematica.* Cambridge: Cambridge University Press.

Wilder, R. L. (1967, February). The Role of the Axiomatic Method. *The American Mathematical Monthly, 74*(2), 115-127. Retrieved 2017, from http://www.jstor.org

Made in the USA
Coppell, TX
28 October 2025

61846277R00098